Realising Farmers' Rights to Crop Genetic Resources

Farmers' Rights are essential for maintaining crop genetic diversity, which constitutes the basis of all food and agricultural production in the world. Realising Farmers' Rights means enabling farmers to maintain and develop crop genetic resources, and rewarding them for their indispensable contribution to the global genetic pool.

This book presents examples of how Farmers' Rights, as they are recognised in the International Treaty on Plant Genetic Resources for Food and Agriculture, are being put into practice in Asia, Africa, Latin America and Europe – in the face of negative trends and huge challenges. Through a series of success stories which include various categories of stakeholders and types of initiatives and policies, it tells of substantial achievements in the four fundamental elements of Farmers' Rights: the rights of farmers to save, use, exchange and sell farm-saved seed; the protection of traditional knowledge; the right to benefit sharing; and the right to participation in decision-making.

The book provides decision-makers and practitioners with practical models and lessons embedded in a conceptual framework of Farmers' Rights. Above all, this book makes clear the urgency of implementing Farmers' Rights – for food security and poverty alleviation, for development and society and for our continued and sustainable existence on Earth.

Regine Andersen is a Senior Research Fellow at the Fridtjof Nansen Institute, Lysaker, Norway.

Tone Winge is a Research Fellow at the Fridtjof Nansen Institute, Lysaker, Norway.

Realising Farmers' Rights to Crop Genetic Resources

Success stories and best practices

**Edited by Regine Andersen
and Tone Winge**

Routledge
Taylor & Francis Group

LONDON AND NEW YORK

from Routledge

First published 2013
by Routledge
2 Park Square, Milton Park, Abingdon, Oxon, OX14 4RN

Simultaneously published in the USA and Canada
by Routledge
711 Third Avenue, New York, NY 10017

Routledge is an imprint of the Taylor & Francis Group, an informa business

British Library Cataloguing in Publication Data
A catalogue record for this book is available from the British Library

Library of Congress Cataloging-in-Publication Data
Realizing farmers' rights to crop genetic resources : success stories
and best practices / edited by Regine Andersen and Tone Winge.
pages cm
Includes bibliographical references and index.
1. Germplasm resources, Plant--Law and legislation. 2. Plants,
Cultivated--Patents. 3. Traditional ecological knowledge--Law and
legislation. 4. Agriculturists--Legal status, laws, etc. I. Andersen, Regine,
1963- editor of compilation. II. Winge, Tone, editor of compilation.
K3876.R43 2013
346.04'69534--dc23
2012050052

ISBN13: 978-0-415-64384-9 (hbk)
ISBN13: 978-0-203-07890-7 (ebk)

Typeset in Bembo by
Fish Books Ltd.

Contents

Figures

Editors and contributors

Editors

Regine Andersen is a political scientist and senior research fellow at the Fridtjof Nansen Institute (FNI) in Norway. She is specialised in research on the effects of international agreements on the management of plant genetic resources for food and agriculture and holds a doctoral degree from the University of Oslo. Her work has focused on the implementation of the International Treaty on Plant Genetic Resources for Food and Agriculture (the Plant Treaty) and in particular its provisions concerning Farmers' Rights. She is the initiator and head of the Farmers' Rights Project, which supports the realisation of Farmers' Rights around the world with research-based guidance (www.farmersrights.org). Dr Andersen has followed the sessions of the Governing Body of the Plant Treaty since its adoption and has organised and facilitated informal international consultations and surveys prior to each of these sessions in preparation of negotiations on the implementation of Farmers' Rights. Major recent publications include *Governing Agrobiodiversity: Plant Genetics and Developing Countries* (Ashgate, 2008) and 'The Plant Treaty – Crop Genetic Diversity and Food Security' (in Andresen, Boasson and Hønneland (eds), *International Environmental Agreements: An Introduction*, Routledge, 2012). Dr Andersen is also the author of numerous publications on Farmers' Rights (available at www.farmersrights.org).

Tone Winge is a research fellow at the FNI. Her research has focused on the implementation of the Plant Treaty, with an emphasis on Farmers' Rights, within the context of the Farmers' Rights Project (www.farmersrights.org). She has also been involved in research on the implementation of the provisions on access to genetic resources and the equitable sharing of the benefits derived from their use in the Convention on Biological Diversity and EU legislation on the marketing of seed and plant propagating material.

Contributing authors

Fetien Abay is an Associate Professor in Plant Breeding and Seed and the Director of the Institute of Environment, Gender and Development Studies at Mekelle University in Tigray, Ethiopia. Her research focuses on genotype by environment interaction, seed systems and participatory plant breeding (PPB). As a gender specialist, Dr Abay has joined numerous evaluation teams in Ethiopia. She is currently involved in a range of international and national research projects on integrated seed sector developments, seed safety through diversity and women and food science. Dr Abay also serves as Associate Editor of the *Journal of Dryland Agriculture,* issued by Mekelle University, and works as a reviewer for various other scientific journals. The special focus of her research and publications has been barley, its diversity, adaptation and role in Ethiopia.

Kamalesh Adhikari is the Research Director of South Asia Watch on Trade, Economics and Environment (SAWTEE) and has more than 12 years of experience of working on trade and intellectual property rights (IPR) policy issues for South Asia, including Nepal. As an expert on IPR and development issues, he has served as an Expert Member of various committees established by the government of Nepal to develop Nepal's policy and legislative frameworks on patents, plant variety protection, biodiversity management, seed production and trade, and food security. He is also a member of the Curriculum Development Committee of the Institute of Agriculture and Animal Science, Rampur. Mr Adhikari has participated as an expert on IPR and development policy in many South Asian and global meetings and has worked with government delegations in preparing Nepal's positions on critical issues under negotiations. He is also the author of several books, book chapters and policy briefs.

Nestor C. Altoveros is an Associate Professor of the Crop Science Cluster at the College of Agriculture, the University of the Philippines, Los Baños. He has worked on plant genetic resources conservation and management since 1976 and collected cereal, vegetable and root-crop germplasm in the Philippines, as well as indigenous vegetable germplasm in Bangladesh, Cambodia, Laos, Malaysia, Thailand and Vietnam. He has represented the Philippines in the negotiations for the Plant Treaty, in several sessions of its Governing Body, as well as at the International Technical Conference on Plant Genetic Resources for the Global Plan of Action in 1996. He also represented Asia in the contact group meeting for the drafting of the Standard Material Transfer Agreement of the Plant Treaty in 2005. He has contributed to the Global Plan of Action for the Conservation and Utilization of Plant Genetic Resources for Food and Agriculture in the Philippines. Among his publications is *The State of the Plant Genetic Resource for Food and Agriculture of the Philippines (1997– 2006)* published by FAO in 2007 (with T. H. Borromeo).

Åsmund Bjørnstad is a Professor of Plant Breeding at the Department of Plant and Environmental Sciences at the Norwegian University of Life Sciences. He has worked with Ethiopian institutions for 20 years and published extensively about breeding issues related to Ethiopian crops. In addition, he has served as academic advisor to ten Ethiopian PhD students.

Teresita H. Borromeo holds an MSc in Plant Breeding and Genetics and a BSc in Agronomy (Plant Breeding) from the University of the Philippines Los Baños. She is a Professor and the Head of the Plant Genetic Resources Division of the College of Agriculture, the University of the Philippines Los Baños, where she teaches crop science, plant breeding and plant genetic resources conservation and management. Ms Borromeo is also a member of the Rice Technical Working Group of the National Seed Industry Council as well as an Examiner for Rice of the Plant Variety Protection Office, Bureau of Plant Industry. As a Philippine delegate, Ms Borromeo has been involved in the negotiations of the Governing Body of the Plant Treaty since 2001. Among her publications is *The State of the Plant Genetic Resource for Food and Agriculture of the Philippines (1997–2006)* published by FAO in 2007 (with N. C. Altoveros).

Salvatore Ceccarelli is a former full Professor in Agricultural Genetics and barley breeder at the International Center for Agricultural Research in the Dry Areas (ICARDA) in Aleppo, Syria. Since 1996 he has been involved in participatory plant breeding programmes, initially in Syria and later in Morocco, Tunisia, Jordan, Yemen, Ethiopia, Eritrea, Algeria and Iran. He is currently working as a consultant on participatory plant breeding in several developing and developed countries.

Robert Chakanda works as an International Seed Specialist and FAO Food Security Adviser in Liberia. Originally from Sierra Leone, he holds a Bachelor's degree in Biological Sciences and Chemistry from the University of Sierra Leone. He has worked with the research team at the West African Rice Development Association (WARDA), conducting research on mangrove rice, and with the Rice Research Station in Rokupr, Sierra Leone, working on sorghum. In 2000, Dr Chakanda graduated from Wageningen University in the Netherlands with an MSc in Plant Breeding. He then went on to work at the Centre for Genetic Resources, the Netherlands, as researcher on Farming Systems, while pursuing another MSc in Geographic Information Science. His PhD was conducted within the work of the Biosystematics Group of Wageningen University and focused on rice genetic resources. Dr Chakanda has travelled widely in Africa and been involved in a range of agricultural research projects. His publications include *Rice Genetic Resources in Post-war Sierra Leone* (2010) and *Farmers' Seed Systems for Sorghum in Mali: An Evaluation of Farmers' Variety Characterization Criteria* (2000).

Mohamed Coulibaly is an environmental lawyer working at the Institute of Research and the Promotion of Alternatives in Development (IRPAD) in Mali. He is Assistant Professor of Environmental Law and International Law at the University of Bamako. Mr Coulibaly works on topics related to sustainable development and has conducted research on the legal, social and environmental aspects of development projects, land policy and other agricultural issues in Mali and elsewhere in Africa. He holds a Masters in International and Comparative Environmental Law from the University of Limoges (France) and a Master of Laws (LL.M.) degree in International Legal Studies from American University Washington College of Law.

Mamadou Goïta is the Executive Director of the Institute of Research and the Promotion of Alternatives in Development (IRPAD) in Mali. He also teaches at the University of Ouagadougou in Burkina Faso; at the National School of Applied Economics, Dakar, Senegal; and at CESAG, the African Centre for Management Studies of Dakar. Mr Goïta has worked on a range of issues including cotton, conflict management, governance, decentralisation, local development and immigration. He has also participated in numerous economic, social and socio-economic studies and assessment processes across Africa. He is a member of the Regional Coordination of the Coalition for the Protection of African Genetic Heritage (COPAGEN) and is active in social movements such as FORAM, the Social Forum of West Africa, the African Social Forum and the World Social Forum. He is the president of several scientific boards and not-for-profit organisations, such as SEXAGON – the Farmers' Union in the Office du Niger (rice producers) and the African Women Economists Network.

Modibo Goïta works for USC Canada in West Africa from the head office in Bamako, Mali.

Stefania Grando is a plant breeder with over 30 years of experience. From 1987 to 2011, she was associated with the barley program of the International Center for Agricultural Research in the Dry Areas (ICARDA) in Aleppo, Syria, and from 2003 as Principal Breeder and Research Manager of the Barley Improvement Section. She has also worked as a Senior Programme Officer at the CGIAR Consortium. She is currently Research Programme Director – Dryland Cereals, at the International Crops Research Institute for Semi-Arid Tropics (ICRISAT).

Helen Groome holds a PhD in Geography from the Autonomous University of Madrid. She worked in the Basque Farmers' Union from 1988 to 2010 and is one of the founding members of the Basque Seed Network. Over the years she has campaigned for GM-free farming, the use of local farmer seeds, farmer seed rights, guaranteed farmer access to farmland and for dynamic and positive relations between farming and the environment, particularly between farming and natural biodiversity. Ms Groome has published numerous articles on these subjects. She is currently working on the Vista Alegre Farm, helping turn the dairy herd organic and setting up and running a dairy to pasteurise milk and make cheeses and yoghurts.

Stef de Haan is an agro-ecologist specialised in crop genetic resource conservation and utilisation. He lives and works in the Andes. He holds a BSc. degree in Tropical Agriculture, an MSc. degree in Agro-ecology, and a PhD in Biosystematics from Wageningen University, the Netherlands. De Haan currently leads the International Potato Center (CIP) Global Program Genetic Resources (GPGR) and is an active member of the NGO Grupo Yanapai. Main interests include *in situ* conservation, breeding, participatory research and nutrition.

Andrew Mushita is the Executive Director of the Community Technology Development Trust (CTDT), Zimbabwe, and has worked for decades with small-scale farmers throughout Southern Africa, in Africa and globally, focusing on seed systems and livelihood options. He often represents the region in international forums on biodiversity and agricultural-related issues, and serves as the Board Chairperson of the Regional Agricultural and Environmental Initiatives Network (RAEIN-Africa) and a board member of the Biotechnology Trust of Zimbabwe and several national committees. He is also a board member of the Association for Plant Breeding for the Benefit of Society (APBREBES). Mr Mushita is active in policy debates on Farmers' Rights, access and benefit sharing (ABS), regional and international agricultural trade issues, and the sustainable use of plant genetic diversity. He has authored and co-authored numerous publications on such issues as food security, environment, biodiversity, ownership and control of seeds, food aid and food sovereignty challenges, Farmers' Rights, implications of agricultural trade and intellectual property rights.

Anitha Ramanna-Pathak is adjunct faculty member at the Symbiosis Institute of International Business, Pune, India. She completed her PhD in International Studies at Jawaharlal Nehru University, New Delhi, and has held a post-doctoral fellowship from the International Food Policy Research Institute, Washington, DC. She was previously a lecturer with the Department of Political Science, Pune University, and Visiting Research Associate at the Indira Gandhi Institute of Development Research, Mumbai. Dr Ramanna-Pathak has been a Fulbright scholar affiliated with the University of California, Berkeley, and Harvard University, and a C R Parkeh fellow at the Asia Research Centre, London School of Economics and Political Science. Her research interests include the politics of development in India, political economy of intellectual property rights, Farmers' Rights, and access and benefit-sharing regimes in developing countries.

Maria Scurrah holds a BA in Biology from Brandeis University, Waltham, Massachusetts, and a PhD in Plant Breeding and Plant Pathology from Cornell University, Ithaca, New York. Born and raised in Huancayo, Peru, Dr Scurrah worked as a breeder at the International Potato Center (CIP) from 1973 to 1989, specialising in nematode and insect resistance. Her work identified rare resistant cultivars in native varieties, which became parental material for several nematode-resistant improved cultivars, in one case through participative selection. From 1990 to 1998 she worked as a plant nematologist at the South Australian Research and Development Institute (SARDI), helping to develop nematode-resistant cultivars. She then returned to CIP as an adjunct scientist while simultaneously working for the NGO Grupo Yanapai, first as its president and later as a coordinator, working on improving smallholder livelihoods, including food security and nutrition, by fostering local agrobiodiversity. She has participated in Farmers' Rights workshops in Zambia, Peru and Ethiopia and is currently living in Lima, Peru.

Pratap Shrestha is a socio-economist by profession, with over 20 years of experience in agricultural research and development in Nepal and South Asia. His work has focused on enhancing the livelihoods of resource-poor farmers through sustainable agriculture, biodiversity conservation and utilisation, and on natural resource management with emphasis on gender and social inclusion, community empowerment and resilience, and sustainable development. Dr Shrestha holds an MSc degree in Agricultural Economics from the University of East Anglia, UK, and a PhD degree in local knowledge and participatory technology development from the University of Wales, UK. He started his professional career in 1990 at Lumle Agricultural Research Center, a DFID (UK)-funded project in Nepal, with various responsibilities from Socio-economist to Head of Planning, Monitoring and Evaluation Unit. He served as Executive Director of Local Initiatives for Biodiversity, Research and Development (LI-BIRD) for five years. Dr Shrestha is currently serving as the Regional Representative and Scientific Advisor for the USC Canada Asia Office in Nepal.

Yoshiaki Nishikawa is a Professor of Agriculture and Resource Economics at the Faculty of Economics, Ryukoku University, Japan. His major research interest is institutional aspects of resource management, especially related to agrobiodiversity. He holds an MSc in Conservation and Utilisation of Plant Genetic Resources and an MSocSc in Development Administration (both University of Birmingham, UK) and a Dr.Agr.Sc. in Global Environmental Economics (University of Tokyo, Japan). He is also actively involved in NGO activities related to seed management by farmers and community development, in African countries and in Japan.

Acknowledgements

This book features the stories of many farmers from Asia, Africa, South America and Europe who have contributed greatly to putting Farmers' Rights in practice. We would like to thank these farmers, as well as all the other stakeholders and decision-makers who have been interviewed, for their time, and for sharing their experiences and reflections. May these examples provide inspiration for all those engaged in the maintenance of crop diversity and the realisation of Farmers' Rights around the world.

In working on this book we have also benefited from the results of the Farmers' Rights Project under the auspices of the Fridtjof Nansen Institute (www.farmers-rights.org); in particular, the informal international consultations and surveys conducted prior to each session of Governing Body of the International Treaty on Plant Genetic Resources for Food and Agriculture. Several hundred stakeholders from diverse sectors all over the world have contributed to shaping the understanding of Farmers' Rights that informs this book. Our thanks go to each and every one of them.

The book has been produced with the support of the Research Council of Norway (RCN Grant 178804/H30), the German Gesellschaft für Internationale Zusammenarbeit (GIZ) and the Development Fund (DF), Norway, for which we are highly grateful. In particular, we would like to thank Ms Annette von Lossau (GIZ), Ms Svanhild-Isabelle Batta Torheim (DF) and Mr Henning Melber (in his capacity as a member of the programme committee of RCN-NORGLOBAL) for moral support and inspiring dialogue.

Mr Tim Hardwick and Ms Ashley Wright at Earthscan/Routledge have been very helpful throughout the process with this book, and we would like to thank both for all their support.

Warm thanks also go to Ms Susan Høivik for excellent language editing and kind support.

Full responsibility for the text and any shortcomings remains, of course, with the editors and authors.

Lysaker, Norway, 3 December 2012
Regine Andersen and Tone Winge

Abbreviations and acronyms

ADCC	Agricultural Development and Conservation Committees (Nepal)
BUCAP	Biodiversity Use and Conservation Asia Programme
CBD	Convention on Biological Diversity
CBDC	Community Biodiversity Development and Conservation Programme
CFPRA	Campagao Farmers' Production and Research Association (Philippines)
CIP	The International Potato Centre/Centro Internacional de la Papa
CNOP/Mali	*Coordination Nationale des Organisations Paysannes du Mali*
COFA	Consumbol Organic Farmers' Association (Philippines)
CSO	civil society organisations
CTDT	Community Technology Development Trust (Zimbabwe)
CVSCAFT	Central Visayas State College of Agriculture, Forestry and Technology (Philippines, now Bohol Island State University)
DADO	District Agriculture Development Office (Nepal)
DUS	distinctness, uniformity and stability
ECOWAS	Economic Community of West African States
EEA	European Economic Area
EHNE	Basque Farmers' Union
FAO	Food and Agriculture Organization of the United Nations
FEDECCH	Federation of Farmer Communities of the Department of Huancavelica/La Federación Departamental de Comunidades Campesinas de Huancavelica (Peru)
GATT	General Agreement on Tariffs and Trade
GCSAR	General Commission for Scientific and Agricultural Research (Syria)
GRBI	Genetic Resources and Biotechnology Institute (Zimbabwe)
HDRA	Henry Doubleday Research Association
ICARDA	International Centre for Agricultural Research in Dry Areas
IER	Institut d'Economie Rurale (Mali)
IPGRI	International Plant Genetic Resources Institute (now Bioversity International)

IPR	intellectual property rights
JSSA	Jaibik Shrot Samrakshan Abhiyan (Nepal)
KMKK	Farmers' Association for Community Development – Riverside (Philippines)
LI-BIRD	Local Initiatives for Biodiversity, Research and Development (Nepal)
LOA	*Loi d'orientation Agricole*
MASFA	Malitbog Sustainable Farmers' Association (Philippines)
NAFOS	National Alliance for Food Security (Nepal)
NARC	Nepal Agricultural Research Council
NGO	non-governmental organisation
NIAS	National Institute of Agrobiological Sciences (Japan)
NordGen	Nordic Genetic Resource Center
Plant Treaty	International Treaty on Plant Genetic Resources for Food and Agriculture
PPB	participatory plant breeding
PRO PUBLIC	Forum for Protection of Public Interest (Nepal)
PVP	plant variety protection
PVSFA	Poblacion Vieja Sustainable Farmers' Association (Philippines)
SAWTEE	South Asia Watch on Trade, Economics and Environment
SEARICE	Southeast Asia Regional Initiatives for Community Empowerment
SoS	Seeds of Survival
SSRs	simple sequence repeats
TRIPS	Agreement on Trade-Related Aspects of Intellectual Property Rights
UMP	Uzumba Maramba Pfungwe (Zimbabwe)
UPOV	International Union for the Protection of New Varieties of Plants
USC	Unitarian Service Committee of Canada
VCU	value for cultivation and use
VDC	Village Development Committee (Nepal)
WTO	World Trade Organization
ZOFRA	Zamora Organic Farmer-Researcher Association (Philippines)

Part I

Introduction

1 Crop genetic diversity and Farmers' Rights

Regine Andersen

This book focuses on an issue of great importance to our continued existence on Earth: Farmers' Rights related to plant genetic resources for food and agriculture. Basically, realising Farmers' Rights means enabling farmers to maintain and develop crop genetic resources, and rewarding them for their indispensable contribution to the global pool, which constitutes the genetic basis of all food and agricultural production in the world. In this book we present a collection of inspiring stories highlighting how Farmers' Rights, as they are recognised in the International Treaty on Plant Genetic Resources for Food and Agriculture (Plant Treaty),[1] are being realised in various parts of the world in the face of negative trends and huge challenges. These stories show how the realisation of Farmers' Rights is crucial not only to the conservation and sustainable use of crop genetic resources but also to farmers' livelihoods. We examine how the various elements of Farmers' Rights can be realised in practice and the potential gains for development and society at large.

This first chapter offers a background and a conceptual framework for understanding Farmers' Rights and what they can mean in practice.

Why Farmers' Rights matter

Plant genetic resources for food and agriculture represent a vast reservoir of fascinating tastes, smells, colours, nutrients – as well as stories and possibilities. An estimated 7,000 species of plants have been cultivated or collected by humans for food (Wilson 1992, p. 275), with often great intraspecies diversity, for some species up to more than 100,000 varieties (FAO 1998, p. 18). This treasure constitutes the genetic basis of all food production. It provides the essential pool from which plant traits can be found that meet the challenges of crop pests and diseases, drought, marginal soils and changing environmental conditions. This is probably more important for farming than any other environmental factor simply because it is what can enable adaptation to shifting environmental conditions – not least, the widely ranging challenges of climate change (Esquinas-Alcázar, 2005; Andersen, 2008; United Nations, 2009; Fujisaka *et al.*, 2010). This also makes it central in the fight against poverty as diversity between and within crops is an effective means of spreading the risks of crop failure for farmers in marginal environments, and access to varieties that can adapt to such conditions is crucial (United Nations, 2009).

The Plant Treaty is aimed at the conservation and sustainable use of crop genetic resources, with the benefits accruing from their use to be shared in a fair and equitable manner. The challenges are considerable, in a world that has lost much of its crop genetic diversity in the course of the past hundred years, and which has still not managed to reverse this negative trend. Whereas *ex situ* conservation (in gene banks, clone archives, etc.) has made substantial achievements over the past decade,[2] far less attention has been focused on *in situ* on-farm conservation and sustainable use. As recognised, and provided for, in the Plant Treaty, both forms are necessary and complementary. Conservation and sustainable use of crop genetic resources *in situ* on-farm is particularly important because it enables the plant varieties to adapt and further develop apace with changing environmental conditions and because the store of knowledge related to the plants is kept alive and further developed when they are in use.

As custodians and developers of genetic diversity in their fields, the farmers of the world are essential in all work for *in situ* on-farm conservation and sustainable use (see Ruiz and Vernooy, eds, 2012). It is important to ensure that farmers can carry on conserving, cultivating and further developing crop genetic diversity – practices that over the past 10,000 years have led to the rich diversity the world has today. Rights in this regard are essential for enabling them to maintain such a vital role for local and global food security. Therefore, the Plant Treaty incorporates specific provisions concerning Farmers' Rights. Their realisation is a precondition for achieving the Plant Treaty objectives of conservation and sustainable use of crop genetic resources, with the ultimate goal of sustainable agriculture and food security (Article 1). Indeed, Farmers' Rights constitute a cornerstone in the Plant Treaty.

Farmers' Rights in the Plant Treaty

Article 9 of the Plant Treaty expresses recognition of the enormous contributions that have been made, and are still being made, by the world's farmers in conserving and sustainably using crop genetic resources, noting that these contributions constitute the foundations for food production around the globe. The Treaty further specifies that responsibility for implementing Farmers' Rights lies with the various national governments. Each country is free to choose the measures deemed necessary and appropriate, in accordance with its own needs and priorities. Although Farmers' Rights are not explicitly defined in Article 9, measures are suggested for protecting and promoting these rights. These include (but are not limited to) the protection of traditional knowledge relevant to crop genetic resources, the right to equitably participate in sharing benefits arising from the use of crop genetic resources and the right to participate in decision-making on the national level on matters related to the conservation and sustainable use of crop genetic resources. Also mentioned are the rights of farmers to save, use, exchange and sell farm-saved seeds and propagating material.

The Plant Treaty contains several other provisions of importance to the implementation of Farmers' Rights. For example, countries are to promote or support (as appropriate) farmers' and local communities' efforts to manage and conserve on-farm their plant genetic resources for food and agriculture and take steps to minimise or if possible eliminate threats to plant genetic resources for food and agriculture (Article 5). Article 6 stipulates that the Contracting Parties shall develop and maintain appropriate policy and legal measures that promote the sustainable use of plant genetic resources for food and agriculture. Various measures are listed for this purpose, among them 'reviewing, and as appropriate, adjusting breeding strategies and regulations concerning variety release and seed distribution' (Article 6.2 [g]). Articles 7 and 8 concern international cooperation and technical assistance with a particular view to strengthening the capabilities of the developing countries to implement the Plant Treaty. The Governing Body of the Treaty, consisting of all Contracting Parties, is to promote the full implementation of the Treaty, including the provision of policy direction and guidance, and monitoring of implementation (Article 19). Article 21 further specifies that the Governing Body is to ensure compliance with all provisions of the Plant Treaty, and the Preamble highlights the importance of promoting Farmers' Rights at the national as well as the international level.

Farmers' Rights: origin of an international-level concept

The history of Farmers' Rights provides an important basis for understanding what they have become today. Let us take a closer look at this history before turning to the contents of these rights.

The idea of Farmers' Rights related to crop genetic resources emerged in the early 1980s as a countermove to the growing and powerful demands for plant breeders' rights, as voiced in international negotiations under the 1983 International Undertaking on Plant Genetic Resources (Mooney, 1983).[3] The purpose was to draw attention to the unremunerated innovations of farmers; innovation seen as the foundation of all modern plant breeding.

In 1987, considerations and practical solutions were suggested under the Commission on Genetic Resources of the United Nations' Food and Agriculture Organization (FAO). These formed a foundation for all further negotiations on Farmers' Rights and provided substantial input to the framing of our current understanding of the concept. The Working Group of the Commission (FAO, 1987) agreed that the breeding of modern commercial plant varieties had been made possible first of all by the constant and joint efforts of all those who had first domesticated wild plants, working to conserve and genetically improve the cultivated varieties over the millennia. Second, thanks were due to the scientists and other professionals who, utilising these varieties as their raw material, had applied modern techniques to achieve the giant strides made in genetic improvements. The Working Group highlighted that some countries had incorporated the rights of plant breeders into plant variety protection laws, to enable them or the commercial companies employing them, to participate in the financial benefits derived from the commercial exploitation of the new varieties. However, there was no explicit acknowledgement of the rights of farmers – no 'Farmers' Rights' as such. The Working Group considered such rights to be fair recognition for the spadework carried out by thousands of generations of farmers; work that had provided the basis for the material available today and to which the new technologies were in large measure applied. The Group agreed that at issue here was not individual farmers or communities of farmers, but the rights of entire peoples who, despite their long-term efforts in breeding, maintaining and improving cultivated plants, had still not shared in the benefits of development.

Opinions differed, however, as to how Farmers' Rights should be recognised, reflecting the deep controversies on control over genetic resources in the 1980s (see Andersen, 2008). Through informal consultations on plant genetic resources known as the Keystone Dialogues, the issue of Farmers' Rights was framed in an instrumental way. That was to lead to the adoption of a resolution by the FAO Conference, for the first time granting formal recognition to *Farmers' Rights*.

FAO Resolution 5/89, adopted by the FAO Conference in 1989, endorsed the concept of Farmers' Rights, specifying that the term refers to rights arising from the past, present and future contributions of farmers in conserving, improving and making available plant genetic resources, particularly those in the centres of origin/diversity. It further stated that these rights are vested in the international community, as trustees of present and future generations of farmers, for the purpose of ensuring full benefits to farmers, and supporting the continuation of their contributions so as to ensure that the need for conservation is globally recognised and that sufficient funds for these purposes will be available. The intention was also to assist farmers and farming communities in the protection and conservation of their plant genetic resources, and of the natural biosphere, and to allow them to participate fully in the benefits derived from the improved use of plant genetic resources, through plant breeding and other scientific methods. This resolution was a milestone – but it was not legally binding.

In 1991 the FAO Conference decided to set up a fund to enable realisation of these rights. Only a few payments were made to the fund, however, so it never really materialised. In 1993, the FAO embarked on lengthy negotiations on what was to become the International Treaty on Plant Genetic Resources for Food and Agriculture, in 2001. Farmers' Rights was among the most contentious issues under the negotiations, and the formulations of Article 9, agreed in 1999, reflect these controversies.

To summarise discussions on Farmers' Rights during these formative years: the main elements concerned balancing the rights of breeders and of farmers, ensuring rewards to farmers for their contribution to the global genetic pool, supporting farmers in conserving and sustainably using crop genetic resources and an international fund to facilitate the funding of such measures. This provides an important background for understanding the contents of Farmers' Rights today.

The contents of Farmers' Rights

With the 2001 Plant Treaty, a legally binding international agreement had been established for the management of plant genetic resources for food and agriculture. States are obliged to protect and promote Farmers' Rights but, as mentioned earlier, are free to choose the measures they deem appropriate, as provided in Article 9.

Although the provisions on Farmers' Rights do not provide a definition of these rights, and the choice of measures is optional, they have provided a basis for establishing a shared understanding of its contents and requirements for realisation. Since 2006, discussions in the Governing Body, as well as through various informal international consultations and surveys involving a broad range of stakeholders from all regions, have contributed to shaping the contents of a common understanding of Farmers' Rights (Andersen, 2005b; Andersen and Berge, 2007; Pistorius *et al.*, 2009; Andersen *et al.*, 2011; Andersen and Winge, 2011).[4] Here we will take the four issues addressed in Article 9, often referred to as the elements of Farmers' Rights, as points of departure in discussing the contents of Farmers' Rights informed by these discussions and consultations. Since Farmers' Rights to save, use, exchange and sell farm-saved seed have been central to all negotiations and discussions on this topic and are basic to farmers' ability to continue their contributions to the global genetic pool, we begin with these. Then we will proceed to the three other main elements: the protection of traditional knowledge, fair and equitable benefit sharing and the right to participate in decision-making.

Farmers' Rights related to farm-saved seed

The Plant Treaty is vague on Farmers' Rights to save, use, exchange and sell farm-saved seed. Section 9.3 of the Treaty states that nothing in Article 9 on Farmers' Rights 'shall be interpreted to limit any rights that farmers have to save, use, exchange and sell farm-saved seed, subject to national law and as appropriate'. This does not offer much direction, except for labelling these practices as 'rights'. The Preamble notes that 'the rights recognised in this Treaty to save, use, exchange and sell farm-saved seed and other propagating material (…) are fundamental to the realisation of Farmers' Rights'. This indicates the importance of these rights, but there is not much guidance here either as the rights referred to are mentioned only in vague terms. Despite the lack of precision, the general line of thought is clear: it is important that farmers be granted rights in this area although the individual countries are free to define the legal space they deem sufficient for farmers regarding their rights to save, use, exchange and sell farm-saved seed.

The question of Farmers' Rights to save, use, exchange and sell farm-saved seed is probably the most contentious in the whole Plant Treaty as it touches on the interests of the commercial seed industry and their prospects for remuneration and making profits while at the same time constituting the ramifications for farmers' ability to conserve and sustainably use crop genetic diversity. In recent decades, the legal space of farmers in this direction has been steadily more restricted in many countries, seriously obstructing farmers' possibilities to continue their crucial work in crop genetic diversity, as we shall see in Chapters 3, 4 and 5 of this book. Important questions concern how Farmers' Rights can be protected and promoted with regard to intellectual property laws (on patents and plant breeders' rights) and seed regulations (on variety release and the marketing of seed and plant propagating material). How can the right balance be struck between the customary rights of farmers and the rights of breeders?

During the preparatory consultations conducted prior to the sessions of the Plant Treaty's Governing Body, a shared understanding among all stakeholder groups emerged: farmers need legal space to continue to perform their role as custodians and developers of crop genetic diversity. Nevertheless, opinion differs as to what this legal space should cover. The challenge is to ensure that farmers can continue their crucial contribution to the conservation and sustainable use of crop genetic diversity to the greatest possible extent while also ensuring that the seed industry gets the revenues necessary for continuing its pivotal work in providing agriculture with the best possible plant varieties. Both are crucial to food security.

Farmers' Rights related to the protection of traditional knowledge

Traditional knowledge is vital for understanding the properties of plants, their uses and cultural significance and how to cultivate them. It also covers practical knowledge of how to select seeds and propagating material, how to store them, use them and manage them for the next harvest. Thus, traditional knowledge is central to the ability of farmers to maintain crop genetic diversity in their fields. Discussions and consultations on Farmers' Rights related to traditional knowledge revealed two approaches to protecting this knowledge: protection against extinction, and protection against misappropriation (see also Andersen, 2013).

Protection against extinction is about ensuring that traditional knowledge is kept alive and can be further developed among farmers. With traditional knowledge disappearing at an alarming pace, along with the genetic erosion in agriculture, measures in this regard are crucial for many farmers engaged in diversity farming. The best way of protecting traditional knowledge against extinction is to use it and share it. Thus, the motto here is *protection by sharing*. There are many ways to promote such sharing. Seminars, field days, farmer field schools, seed fairs and other gatherings can provide useful and often inspiring arenas to share knowledge. Documentation in seed catalogues, registries, books, magazines, on videos and web sites (for example, of gene banks) constitute further useful possibilities. Oral and written ways of sharing knowledge are complementary, and the combination of both constitutes the optimal way of ensuring that the knowledge can be kept alive, shared and used.

Protection against misappropriation is a different approach. It is based on the possibility that farmers' varieties together with associated knowledge may be 'discovered' and developed by commercial actors, exclusively monopolised by use of intellectual property rights – and without benefit-sharing mechanisms. This is seen as a possibility especially as regards varieties that have not been documented and thus cannot be proved to be *prior art* (formal knowledge of already existing plant varieties – which is a necessary formal criterion for establishing whether a variety for which plant breeders' rights are sought is really 'new'). According to this approach, sharing of knowledge should not take place unless there are measures in place to avoid misappropriation. Documentation of various kinds to establish *prior art* is highly useful to avoid misappropriation of crop genetic material; also important are legal measures, like regulations on bioprospecting and user-country measures as provided for in the Convention on Biological Diversity (CBD) and conditions for intellectual property rights as discussed *inter alia* under the World Trade Organization Agreement on Trade-Related Aspects of Intellectual Property Rights (TRIPS, see Andersen, 2008).

How great are the risks of misappropriation of farmers' varieties and knowledge? Are the actual risks worth the measures taken to avoid misappropriation as well as the fears it creates among farmers – in a cost-benefit perspective? So far there are very few known examples of misappropriation of farmers' knowledge related to crop genetic resources. The challenge is to balance these concerns in such a way that traditional knowledge can still be shared to the greatest extent possible. In the 2010 Global Consultations on Farmers' Rights (Andersen *et al.*, 2011, Andersen and Winge, 2011), the great majority of participants noted that protection against extinction is the most important concern as regards the protection of traditional knowledge.

Farmers' Rights related to equitable benefit sharing

The next measure for protecting and promoting Farmers' Rights, as indicated in the Plant Treaty, concerns the right to participate equitably in the sharing of benefits arising from the utilisation of plant genetic resources for food and agriculture (Article 9.2 [b]). The Treaty provides no further details as to what this might mean in practice. However, Article 18 on the Multilateral System on Access and Benefit Sharing lists the most important benefits as follows: (1) facilitated access to plant genetic resources for food and agriculture; (2) the exchange of information; (3) access to and transfer of technology; (4) ; and (5) the sharing of monetary and other benefits arising from commercialisation. Moreover, it is specified that benefits arising from the use of plant genetic resources for food and agriculture that are shared under the Multilateral System should flow primarily, directly and indirectly, to farmers in all countries but especially in developing countries and countries with economies in transition, who conserve and sustainably utilise plant genetic resources for food and agriculture.

Whereas these provisions all relate to the Multilateral System and not directly to the provisions on Farmers' Rights in the International Treaty, they reflect a line of thought on benefit sharing which is relevant for interpreting Article 9.2 (b) on benefit sharing as a measure to protect and promote Farmers' Rights. First, we note that there are many forms of benefit sharing, with monetary benefits comprising only one part. Second, benefits are not to be shared only with those few farmers who happen to have plant varieties that are utilised by commercial breeding companies but with farmers everywhere who are engaged in the conservation and sustainable use of crop genetic diversity.

Here we see the thinking from the early days of FAO negotiations on Farmers' Rights: benefits should be shared between 'entire peoples', the stewards and developers of plant genetic resources in agriculture and society at large. The underlying reasoning is that it is their legitimate right to be rewarded for their contributions to the global genetic pool, from which we all benefit; further, that it is an obligation of the international community to ensure such recognition and reward. Examples of benefit-sharing measures include access to seeds and propagating material and related information for farmers; participatory plant breeding in collaboration between farmers and scientists; strengthening of farmers' seed systems; conservation activities, including local gene banks; and enhanced utilisation of farmers' varieties, including value adding and market access. In practice, most such benefit-sharing measures today take place in collaboration between farmers and scientists/plant breeders or between farmers and non-governmental organisations (NGOs). The Benefit Sharing Fund of the Plant Treaty has increasingly contributed to benefit-sharing projects. Furthermore, multilateral, bilateral and non-governmental development cooperation agencies have been contributing to projects in developing countries.

Farmers' Rights related to participation in decision-making

Article 9.2 (c) concerns the right to participate in making decisions at the national level, on matters related to the conservation and sustainable use of plant genetic resources for food and agriculture. To understand what this measure might mean in practice, we need to specify the relevant matters in which farmers could have the right to participate. Also the forms of participation need specification. Further important questions concern which farmers should participate and how they could be represented.

Relevant matters for farmers' participation are decision-making related to laws, regulations, policies and programmes affecting the management of plant genetic diversity in agriculture – in particular, laws and regulations on variety release and the marketing of seed and propagating material; plant variety protection laws; patent laws; bioprospecting laws or regulations; laws on conservation and sustainable use of biodiversity and crop genetic resources; and legislation on the rights of indigenous peoples and traditional knowledge. Ideally, policies and programmes targeted at farmers should take as their point of departure the real-life situations and perspectives of farmers, based on farmer participation. Also relevant is agricultural legislation that is not aimed at regulating the management of crop genetic diversity but nevertheless may affect such management, as well as Farmers' Rights. The implementation of laws and regulations is also relevant to farmers' participation, as well as the implementation of programmes on crop genetic diversity. The ways in which these are interpreted and implemented may be important to the effects on farmer management of these resources and thereby also on their very livelihoods. Furthermore, decisions concerning the operations of gene banks, genetic resource centres, public or publicly supported plant breeding institutions and other institutions aimed at the management of crop genetic diversity are relevant for participation from farmers.

Extensive use of public hearings at various stages in the decision-making process is an important measure for ensuring participation. It is essential that farmers engaged in the management of plant genetic diversity are aware of the processes and are explicitly invited to participate through their organisations. Farmers and their organisations may also take the initiative to seminars and meetings as well as contribute to newspapers and other media channels in order to seek influence on the decisions, and they may carry out advocacy work by other means as well. As for the implementation of laws and regulations, programmes and the operations of relevant institutions, boards, reference groups or the like are normally established for control or guidance. Farmers' representation and participation in such bodies should be a central concern.

'Farmers' are not defined in the Plant Treaty. However, in several provisions, the Treaty emphasises those who conserve and develop plant genetic resources for food and agriculture. This might indicate that such farmers are the core target group to the provision on farmers' participation in decision-making at the national level concerning crop genetic resources. That, however, should not be taken to mean that farmers' participation is limited to this group.

In some countries, most farmers belong to farmers' organisations; in other countries few farmers are involved. However, even in countries with high membership levels it may be difficult to identify legitimate representatives of farmers because not all organisations have participatory and democratic methods of electing their representatives. Farmers specifically involved in the conservation and sustainable use of crop genetic diversity are often not organised as such. However, in many countries NGOs are involved in the realisation of Farmers' Rights. Such organisations may have significant experience and insight in the situations and needs of farmers. They may be in a position to initiate processes among farmers to select farmer representatives – or the NGOs themselves may identify farmer representatives although ideally farmers should be the ones to select their own representatives. In some situations it could be relevant to consult with NGO representatives in decision-making processes related to crop genetic resources with a view to farmer perspectives.

There are two important preconditions for greater participation of farmers in decision-making. First, decision-makers need to be aware of the vital role played by farmers in conserving and developing plant genetic resources for food and agriculture in order to understand why their participation is required. Second, farmers may need additional support and capacity building to be in a position to participate effectively in complex decision-making processes. Awareness-raising and capacity building are thus central measures for enabling farmers to become involved in the making of decisions that affect them, their work – and thereby all of us.

Notes

1   The International Treaty on Plant Genetic Resources for Food and Agriculture was adopted in 2001 and entered into force in 2004. It is the first legally binding international agreement on the management of plant genetic resources for food and agriculture and is aimed at the conservation and sustainable use of these resources and the fair and equitable sharing of the benefits arising from their utilisation (for more information, see www.planttreaty.org). For in-depth analyses of challenges related to the implementation of the Plant Treaty, see Halewood *et al.* (eds), 2013 and Andersen, 2008.

2   With the Global Crop Diversity Trust (www.croptrust.org) and the Svalbard Global Seed Vault (www.regjeringen.no/en/dep/lmd/campain/svalbard-global-seed-vault.html ?id=462220), important instruments are in place which support *ex situ* conservation all over the world with financial resources and safe back-up storage facilities.

3   This section is based on Andersen, 2005a.

4   The process is documented at www.farmersrights.org/about/fr_in_itpgrfa.html

2 What constitutes a success story in the context of Farmers' Rights?

Regine Andersen and Tone Winge

The realisation of Farmers' Rights is already taking place, at various levels and in different parts of the world. Despite barriers and challenges, we can find many examples of initiatives, projects, legislation and policies, which contribute to the realisation of Farmers' Rights and can help to provide models and important lessons for further work. In this book we have collected success stories and best practices. In the present chapter we explain what we mean by *success stories,* and how the concept can be used with regard to the realisation of Farmers' Rights. We present the various success stories that constitute the main body of the book and end this chapter with an overview of the main elements of each story.

'Success story' as a concept related to Farmers' Rights

By *success stories* we mean policies, measures, projects or activities that have resulted in substantial achievements with regard to one or more of the elements of Farmers' Rights as set out in the Plant Treaty.[1] These policies, measures, projects or activities are not necessarily 'perfect': problems or challenges encountered on the way can also stand as lessons from which others can learn. The main criterion is that significant achievements have been made and that these can provide inspiration and lessons for others.

Achievements can be made at different levels. There can be intermediate achievements on the way towards a larger goal. In this sense, not only achievements of ultimate goals are relevant as success stories in our context. Also the smaller steps on the way to a goal – reaching partial goals of various kinds – can be seen as significant achievements that can inspire and motivate other stakeholders to take further steps. In this book we have gathered stories of smaller and greater successes, to display the wide range of achievements already being made on the path to the wider realisation of Farmers' Rights.

When the elements of Farmers' Rights in the Plant Treaty are taken as the point of departure for identifying success stories, what does this mean in operational terms? Among what kind of policies, measures, projects and activities could we identify achievements in the realisation of Farmers' Rights? On the basis of the deliberations presented in Chapter 1, we have looked for policies, measures, projects and activities for this collection of success stories that:

- contribute to enabling farmers to save, use, exchange and sell farm-saved seed;
- aim at documenting and sharing traditional knowledge among farmers in order to avoid loss of such knowledge, or to protect farmers' traditional knowledge against misappropriation while also ensuring that such knowledge can be shared;
- promote equitable benefit sharing – such as funding mechanisms that support farmers in conserving and sustainably using plant genetic resources; participatory plant breeding projects resulting in added value to farmers' varieties; community gene banks used effectively in farmers' breeding or farming strategies; marketing strategies to enhance the demand for diverse crop products; other incentive structures to motivate conservation and sustainable use of genetic resources; recognition of farmers' contributions, for example in the form of awards;
- strengthen farmers' participation in decision-making, for example by involving farmers in national consultative processes related to the management of crop genetic resources, breeding strategies or more specifically to Farmers' Rights; capacity-building activities that lead to greater involvement of farmers in relevant decision-making; or advocacy by farmers' organisations aimed at improving policies on genetic resources and Farmers' Rights. Also, awareness-raising of the important role played by farmers in conserving and developing PGRFA is relevant here.

However, in 'real life', policies, measures, projects and activities often contribute to more than one of the elements of Farmers' Rights. In the present volume, we have categorised each success story according to the element within which they made the greatest achievement. In addition, we also show how they contributed to other elements of Farmers' Rights. In the following, we explore in greater detail what 'success' can involve within the four categories.

What are successes regarding Farmers' Rights to save, use exchange and sell farm-saved seed?

The possibilities that countries have to define legal space for farmers with regard to their rights to save, use exchange and sell farm-saved seed are largely framed by their international commitments. Most countries in the world are now members of the World Trade Organization (WTO) and are thus obliged to implement the WTO Agreement on Trade Related Aspects of Intellectual Property Rights (TRIPS). The TRIPS Agreement states that all WTO member countries must protect plant varieties either by patents or by an effective *sui generis* system (a system of its own kind) or a combination, although the text does not explicitly define the limits to a *sui generis* system or the meaning of an 'effective' *sui generis* system (Article 27.3[b]). At any rate, we can say that countries are supposed to introduce some sort of plant breeders' rights in order to compensate plant breeders' for their contribution and to stimulate further innovations in plant breeding.

The International Union for the Protection of New Varieties of Plants (UPOV) has held that the most effective way to comply with the provision of an effective *sui generis* system is to follow the model of the UPOV Convention, and there are several proponents of this stand. The UPOV model exists in several versions: according to the most recent one (the 1991 Act of the UPOV Convention), plant breeders are to be granted comprehensive rights over protected varieties – to the detriment of farmers' customary rights to save, re-use, exchange and sell seeds. It is possible to grant exemptions for small-scale farmers to enable them to save and re-use seeds but only within strict limits. Exchange and sale of seeds among farmers is prohibited. All this applies to seeds of varieties protected with plant breeders' rights – not to seeds of landraces and other farmers' varieties that are not covered by such rights. As chapters 3 and 14 of this volume show, following UPOV '91 is not a requirement to comply with the TRIPS Agreement. The TRIPS Agreement provides only minimum standards, leaving enough scope for the development of other solutions more compatible with the demand for Farmers' Rights (see e.g. CIPR, 2002; Helfer, 2002; Correa, 1998).

WTO member countries are faced with the challenge of meeting their TRIPS obligations regarding plant breeders' rights while also creating the necessary legal space for the realisation of Farmers' Rights under the Plant Treaty (Andersen, 2008; Santili, 2012). Also, regional and bilateral trade agreements set the introduction of plant breeders' rights as a condition in many cases. Such regimes are evolving rapidly in many countries, increasingly restricting the legal space available to farmers. Moreover, in some cases the national seed sector never had the chance to adapt to a slowly developing intellectual property regime. This makes it difficult to establish *prior art* (formal knowledge of already existing plant varieties), which is necessary to identify whether a variety for which plant breeders' rights are sought is really new.

An additional constraint to Farmers' Rights in many countries, is legislation on plant variety release and the marketing of seeds and plant propagating material requiring that plant varieties must meet certain criteria to be approved for marketing and that the seeds are formally certified. The background is the need to ensure plant health and seed quality. Nevertheless, recent years have seen a development towards stricter regulations also here, which impinge on Farmers' Rights to save, use, exchange and sell farm-saved seed. Such regulations affect not only varieties protected by intellectual property rights but also those that are not protected, like landraces and farmer varieties of particular importance to those farmers who engage in the conservation and sustainable use of crop genetic resources. As landraces and farmers' varieties are often not genetically homogeneous enough to meet the requirements of such regulations, the result is that these varieties are excluded from the market. Efforts to loosen these strict regimes in the European Union have contributed to some more legal space for the marketing of what are termed 'conservation varieties', as we shall see in Chapter 4 – but they do not solve the problems. Often such legislation also stipulates that only authorised seed shops are allowed to sell seeds, and farmers are not allowed to market or exchange such seed. Such regulations, together with strict plant breeders' rights, represent a serious obstacle to the traditional rights of farmers to save, use, exchange and sell seed.

In addition, patents enabling the protection of plant properties or breeding processes constitute an emerging challenge with regard to farmers' legal space to save, use, exchange and sell farm-saved seed. Such patents provide exclusive rights to the rights-holders, which are far stricter than plant breeders' rights.

What constitutes a success story with regard to ensuring Farmers' Rights to save, use, exchange and sell farm-saved seed in light of these challenges?

In countries with legislation on plant variety protection, variety release and seed marketing, limiting the rights of farmers to save, use, exchange and sell farm-saved seed, a positive achievement (or 'success story') can involve making a regulation less stringent, or preventing a stricter regulation from being adopted. This has happened in Norway as shown in Chapter 3. This story tells how the Norwegian government in 2005 decided to reject a bill on plant breeders' rights on the grounds that it would limit the rights of farmers. It also tells how the regulations on variety release and seed marketing were amended in 2010 to provide Norway's farmers with more extensive rights to save, use, exchange and sell farm-saved seed. We will see why and how these decisions were made and derive some possible lessons for other countries.

Where regulations are very strict and there seems little scope for achieving legal changes, the question is how to proceed. Are there possibilities to enable farmers to save, use, exchange and sell farm-saved seeds despite existing laws? Developing such a possibility would constitute an achievement. The story from Spain's Basque Country in Chapter 4 provides an example where precisely such possibilities have been identified and put into practice. This is the story of the Basque Seed Network and how it has succeeded in conserving local varieties through documentation and use, and in the face of detrimental laws and a lack of legal space for the traditional practice of seed sharing. This chapter shows how even relatively small-scale initiatives can have an impact on the conservation of crop genetic resources and the realisation of Farmers' Rights and discusses possible lessons for other countries.

The ultimate achievement, if seen solely from the perspective of Farmers' Rights, would be to grant farmers the rights to freely save, use, exchange and sell farm-saved seed under intellectual property law as well as under seed regulations. Other solutions would be needed in terms of compensation and incentives to plant breeders and to deal with concerns relating to plant health and seed quality. India stands out as the country with the most far-reaching legislation on the protection of plant varieties and farmers rights in the world and the country that comes closest to such an ultimate achievement. Under India's Protection of Plant Varieties and Farmers' Rights Act of 2001, farmers are entitled to freely save, use, exchange and sell farm-saved seed: only they may not sell seeds of protected varieties under their brands. This act provides farmers with other extensive rights as well. Chapter 5 tells the story of how this law came to be passed and discusses its impact and possible lessons for other countries.

What are successes regarding traditional knowledge related to crop genetic diversity?

One measure to protect and promote Farmers' Rights, as set out in Article 9.2 (a) of the Plant Treaty, involves the protection of traditional knowledge relevant to plant genetic resources for food and agriculture. However, the Treaty fails to specify this in greater detail.

At the informal international consultations on Farmers' Rights held in Lusaka in 2007, various examples were given and proposals offered on how national or local governments could support such initiatives (Andersen and Berge, 2007). Ideally, farmers' varieties and associated knowledge should be documented and seeds stored in gene banks, in order to ensure that these valuable resources are shared and do not become extinct. Also, measures to keep the knowledge alive and sharing it 'live and active' were emphasised. However, several participants expressed concern about the legal status of gene-bank collections. If seeds are readily available, they might also be picked up by commercial actors and used without obtaining prior informed consent from the farmers or benefit-sharing arrangements. There was widespread concern that local communities might lose control of their plant genetic resources, particularly if modified forms of these resources were then made subject to intellectual property rights. On the other hand, there have been very few actual examples of such misappropriation related to crop genetic resources. This situation points to the difficult dilemma between sharing seeds and traditional knowledge to avoid extinction – and protecting it against misappropriation.

Participants at the Lusaka consultation also expressed regret that, due to fears of misappropriation, it was felt necessary to show such caution in many regions with activities so vital for further availability of genetic resources and related knowledge. Such fears obstruct conservation work aimed at enhancing farmers' varieties and strengthening their seed systems – which is crucial to the future of our plant genetic heritage.

In light of these considerations, an ultimate goal for policies, measures, projects and activities aimed at protecting traditional knowledge related to crop genetic diversity would be to facilitate documentation and free sharing of such knowledge among farmers – while also ensuring that no misappropriation can take place.

One challenge in registering and documenting traditional varieties of plants lies in the genetic heterogeneity of these varieties. Many are difficult to describe as varieties and that is part of the problem when it comes to the fear of misappropriation. For a plant breeder to be granted plant variety protection, it is in countries with *inter alia* UPOV '91-compatible legislation sufficient to discover a variety and develop it, for example in terms of genetic purification. If the prior existence of the variety cannot be documented, farmers will often not be in a position to challenge such a right. For that reason, developing improved methods of documenting traditional varieties can represent important achievements in protecting traditional knowledge against misappropriation – as well as against extinction, as we will see in an example from Peru.

Chapter 6 tells the story of a catalogue documenting the extensive potato varieties and traditional knowledge of Quechua farmers in Peru's Huancavelica region: a project that involved eight communities and took three years. The story highlights the impact of the catalogue and how involved stakeholders viewed it a few years after publication. Cataloguing can serve as an important means for maintaining traditional knowledge and protecting it against misappropriation, as this story suggests.

A community rice seed registry in Bohol in the Philippines provides a further example in this regard, as reported in Chapter 7. While the methods of registration are not as sophisticated as in the Peruvian case, the seeds are stored in a field gene bank in collaboration with a local university and are made accessible to all interested farmers. The farmers who are involved in the registry invest great effort in registering all varieties in use, together with the related knowledge. As it turned out, there have been no cases of alleged misappropriation of any of the varieties in the registry. Nevertheless, the seed registry is seen as providing security for the farmers to share seeds and knowledge as they wish, without having to worry about misappropriation. An important measure here was securing recognition of the registry from the local authorities. This recognition has fostered greater self-esteem with regard to the involved farmers' shared knowledge and their contribution to the local genetic pool – an important achievement in itself.

What are successes regarding equitable benefit sharing?

As shown in Chapter 1, the Plant Treaty provides several points of departure for an understanding of what farmers' participation may entail as to the equitable sharing of benefits generated from the use of crop genetic resources, as per Article 9.2 (b). Success may be measured in various ways. One could focus on the parties involved in the sharing of benefits and seek to identify best practices among them. Here it would be relevant to distinguish between funding parties (e.g. the Benefit Sharing Fund under the Plant Treaty, donor agencies and NGOs and national authorities), and the parties involved in practical benefit sharing, like plant breeders, scientists, extension services and local NGOs on the one hand – and various forms of farmer groups and organisations on the other.

Another approach could be in terms of the measures needed to compensate and recognise farmers for their contribution to the genetic pool. Such an approach could focus on various forms of *incentive structures,* like subventions and loans on favourable conditions, grants for areas cultivated with certain crop varieties of value for conservation and sustainable use, facilitation of the marketing of relevant products and other infrastructure measures; *direct support* to projects and initiatives aimed at conservation, development and sustainable use of crop genetic resources among farmers; and *rewards* to farmers for outstanding contributions.

As most cases of benefit sharing to date have been related to direct support to projects and initiatives aimed at the conservation, development and sustainable use of crop genetic resources among farmers, this is the focus of the present book and will serve as our point of departure for what constitutes 'success'.

Our first story of benefit sharing is from Syria (Chapter 8). It tells about a participatory plant breeding project that was led by the International Centre for Agricultural Research in Dry Areas (ICARDA) and for a time encompassed 24 villages. The village of Kherbet El Dieb is used as an example, but stories and farmers' views from other participating villages are also featured. The project has resulted in varieties of barley that perform very well in local climates, with substantial increases in yield. Whereas this is an achievement in itself, the really interesting part is how it was reached, with farmers and scientists working together to adapt the varieties directly to the fields in question. A related achievement was the effect this collaboration had on the participating farmers in terms of self-esteem and capacity building. As such, it is a success story of benefit sharing between farmers and breeders to enhance and further develop crop genetic diversity adapted to the particular environmental conditions and consumer needs of the farming communities.

In Nepal, the NGO LI-BIRD has been working with various farming communities to conserve and further develop local plant genetic resources, and to ensure that the communities benefit from maintaining this diversity. Chapter 9 tells the story of what happened when farmers in the hills around Pokhara in central Nepal came to realise the value of the crops in their fields. We meet Ms Maina Thapa, a member of the Pratigya Cooperative, who has greatly improved her livelihood. Her success was possible through a project that helped in identifying valuable and endangered crop varieties and establishing the infrastructure and measures required to bring them to urban markets. We also meet a famous farmer breeder, Mr Surya Adhikari, who has had great success in breeding drought-resistant rice varieties. This is a success story of benefit sharing where a national NGO, supported by international donors and organizations, played an important role. The important achievement here is the improved livelihoods resulting from the discovery, growing and marketing of traditional varieties that had almost disappeared. It is also a major achievement that farmers, with the help of LI-BIRD, are in a position to breed rice well-adapted to the drier climate conditions already emerging in the Himalayan regions.

Seed fairs are an inspiring way to provide farmers with greater access to crop genetic diversity. They allow farmers to get a good overview of available varieties and share experiences and knowledge, as well as providing arenas for meetings and gatherings that can foster capacity building and empowerment. Chapter 10 tells about community seed fairs held in the Uzumba Maramba Pfungwe District in Zimbabwe, assisted by the Community Technology Development Trust (CTDT). This is a success story of benefit sharing between the CTDT and local farmers; the major achievement has been the substantially improved access to seed and related knowledge in the district, in turn resulting in better livelihoods for the people.

In Chapter 11 we see how traditional knowledge, variety selection and farmer innovation still in use can be supported and further enhanced through farmer–scientist collaboration. This story comes from the Tigray region of Ethiopia and highlights the achievements of Mr Kahsay Negash, a 92-year-old farmer, and the varieties he has developed on the basis of local landraces of barley. It is a success story of benefit sharing between Ethiopia's Mekelle University (and its external donors) and the local communities. The major achievement has been the further development of local varieties to new and better varieties, in turn leading to increased cultivation of local crop diversity – so that the variety traits and associated knowledge of the original local varieties have been maintained.

From Mali we provide a similar story in Chapter 12. Here the Seeds of Survival (SoS) project was carried out in partnership between a Canadian NGO (USC Canada) and local partners, including local-level farmers' organisations, local governments, local associations and NGOs and local administrative authorities. Major achievements were made in terms of capacity building, farmer involvement and stakeholder collaboration to re-introduce traditional varieties, improve their conservation and increase production.

Finally in Chapter 13, we highlight a story from Japan, where the Hiroshima Agricultural Gene Bank re-introduced local varieties along with traditional knowledge. This success story illustrates the close link that often exists between the elements of Farmers' Rights, in this case the protection of traditional knowledge and equitable benefit sharing. The major achievement has been the contribution to increased use of local plant genetic diversity and traditional knowledge.

What are successes regarding participation in decision-making?

The fourth measure to protect and promote Farmers' Rights as suggested in the Plant Treaty, concerns the right to participate in making decisions at the national level, on matters related to the conservation and sustainable use of plant genetic resources for food and agriculture (Article 9.2 [c]). As discussed in Chapter 1, farmers' participation in decision-making is relevant with regard to legislation, regulations, policies and programmes aimed at, or affecting, the conservation and sustainable use of crop genetic diversity, as well as the implementation of these. Also relevant is decision-making regarding the operations of institutions dealing with crop genetic diversity. Consultative processes of various kinds are central, and the better represented farmers are, the greater legitimacy the results will have and the more likely it is that they will constitute effective measures for the conservation and sustainable use of crop genetic resources as well as the realisation of Farmers' Rights. Finally, awareness-raising and capacity building among farmers and decision-makers alike are important for ensuring participation in decision-making.

The ultimate 'success story' of farmers' participation in decision-making at the national level would be a story of a comprehensive consultative process to mainstream, harmonise and improve legislation, policies and programmes towards the conservation and sustainable use of crop genetic resources and the realisation of Farmers' Rights. So far, the only example of such a process known to the authors is the on-going development of a plan for the realisation of Farmers' Rights in Norway, which is aimed at achieving precisely that. As this process is in its infancy, it is still too early to tell this story in full, with or without a happy ending.

However, there are several examples of consultative processes related to individual acts of legislation or specific regulations. In particular in Europe and North America, consultation processes are quite common, but there are also traditions of consultations in many developing countries. In Chapter 5 we explain how India's Protection of Plant Varieties and Farmers' Rights Act of 2001 was developed through an extensive consultation process. Also, the story in Chapter 4 of how Norway developed its plant variety and seed legislation is an example of extensive consultative processes. Whereas these stories are presented in the context of Farmers' Rights to save, use, exchange and sell farm-saved seed, they are also relevant in the context of farmers' participation in decision-making.

Farmer representation on boards and in reference groups of institutions relevant to the management of crop genetic resources is not yet very common, nor is awareness-raising among decision-makers on the importance of farmer participation. Here much remains to be done. However, we can find cases of awareness-raising and capacity building among farmers, enabling them to participate in consultation processes and decision-making. Many of the chapters in this book offer proof of that.

In Chapter 14 we turn to one success story that exemplifies farmers' participation in decision-making at the national level. We see how NGOs in Nepal, through an alliance and in close consultations with farmers, influenced the decision-making regarding plant variety protection legislation in the country. This story offers valuable lessons on how organisations involved in the maintenance of crop genetic resources and the promotion of Farmers' Rights can influence official policies through networking, collaboration and communication, within civil society and with the government and the media. It is the story of a highly successful advocacy campaign.

Key elements of the success stories

The success stories of this volume are meant to inform about the possibilities and potentials of realising Farmers' Rights, and to provide inspiration for further initiatives in this regard. Therefore they are presented in considerable detail and with certain common features. Particular attention has been paid to actors involved; the political and legal framework conditions; a detailed history of how it all evolved; the crop genetic diversity involved and geographical outreach; the number and ways in which farmers were involved; how the results have affected the management of crop genetic resources and farmers' livelihoods; and problems and challenges encountered on the way. It has been important to highlight the main achievements and conditions for success – as a basis for lessons and inspiration for others. Throughout this book, it is the actual experiences of real-life farming communities that are central to the stories we wish to share with you.

Note

1   The understanding of what constitutes success with regard to the realisation of Farmers' Rights presented in this chapter is based on Andersen and Winge (2008) and develops this approach further.

Part II

Success stories from the realisation of Farmers' Rights to save, use, exchange and sell farm-saved seed

3 Norway's path to ensuring Farmers' Rights in the European context

Regine Andersen

Norway has maintained a high profile in international efforts to maintain crop genetic diversity – as a driving force in the negotiations leading up to the Plant Treaty; as a bridge-builder between North and South; as a financial contributor to international processes and tasks; and not least, by realising the Svalbard Global Seed Vault. But how are things on the home front?[1]

Very few Norwegian landraces and farmers' varieties of cereals, potatoes and vegetables have survived (Andersen, 2012). As to fruit and berries, the picture is somewhat brighter, but plant breeding on this material has been very limited, so this diversity is also under threat. In former days Norway boasted a wide range of local meadow plants, but also here, much has disappeared. Beyond doubt, the modernisation of agriculture has meant greater production efficiency – but it has also led to widespread genetic erosion, in Norway as elsewhere. Although much crop genetic diversity has been lost in Norway, substantial efforts are being made to save what is left through the Nordic Genetic Resource Center (NordGen) established by the Nordic countries and through the Norwegian Genetic Resource Centre.

The main challenges to farmers' contributions to the genetic pool in Norway relate to formal regulations on Farmers' Rights to save, use, exchange and sell farm-saved seed. Nevertheless, Norwegian farmers are far better off than their colleagues elsewhere in Europe – thanks not least to the importance that the Norwegian government has placed on Farmers' Rights. In this chapter we begin by looking at a proposal that was aimed at strengthening the regulations on plant variety protection and was rejected by the Norwegian government on the grounds of Farmers' Rights – a success story in itself. Then we turn to Norway's current regulations concerning plant variety release and seed marketing, which represent a considerable improvement from the situation only a few years ago. With these two success stories we do not mean to say that all problems have been solved. Seed regulations still constitute great hurdles for farmers engaged in crop genetic diversity in Norway. Nevertheless, as compared to the previous situation and in light of Norway's international commitments – in particular under the European Economic Area Agreement, which associates Norway to the European Union without entailing full membership – these examples can indeed be seen as success stories.

Norway – a country with special conditions

Norway represents a small and very distinctive market for seed and propagating material. This is partly because the agricultural sector is small and still shrinking (only 3.2 per cent of the total area is presently used for agriculture due to inhospitable terrain, and the number of work years in agriculture is only around 50,000), but it is also because the country lies so far north. Temperatures are at times comparable to those of several other countries, but light conditions are very different. Summers are short but with many hours of daylight, extremely short nights and great changes in light conditions over the course of the growing season. It is therefore essential for plants to be able to react not only to temperature but also to light conditions when they prepare for winter. Moreover, plants not adapted to high-latitude light conditions will often experience stress because of the lack of rest periods during the short summer nights and may fail to thrive.

This rather uncommon situation, which Norway shares with very few other countries, makes the seed/propagating materials market less attractive for multinational and other foreign companies, and much of the seed sector is still Norwegian-owned. An exception is grass seed, where almost everything is imported. Another exception concerns berries, where there has been little breeding or development in Norway aside from strawberries and raspberries. Vegetable varieties are also largely imported and tested for adaptability to Norwegian conditions before they are made available for sale. Most of these varieties are hybrids, which means that farmers will normally have to buy new seed every year. Today's situation as regards vegetables and berries is the result of – *inter alia* – a lack of resources (and thus capacity) for plant breeding in a small country with such distinctive needs. Plant breeding is an important factor in the conservation and sustainable use of crop genetic diversity. Thus, Norway seeks to balance Farmers' Rights *and* breeders' rights.

Most of Norway's farmers are satisfied with the available assortment of seeds and propagating material from commercial varieties – but many organic farmers find that their needs are not met by supplies available from authorised seed shops and seek out other channels. It is among these farmers that we find the majority of what may be termed biodiversity farmers[2] in Norway today: there are not many of them, perhaps not more than a few hundred. In this perspective, the issue of Farmers' Rights becomes particularly important, as it is all about enabling these relatively few farmers to continue their work to broaden the genetic base of crops as well as the diversity of crop varieties adapted to the conditions and demand in Norway. Promoting Farmers' Rights may also encourage other farmers to contribute to the conservation and sustainable use of crop genetic diversity.

Norway's rejection of stricter plant breeders' rights

Norwegian legislation on plant breeders' rights was adopted in 1993 and led to membership in the International Union for the Protection of New Varieties of Plants (UPOV).[3] While a few changes have been made to the law since then, they have been insignificant. This has meant that Norway's farmers are entitled to save seed from their own harvest of protected varieties for use the following season. The law

Figure 3.1 The 2011 Norwegian Plant Heritage Award winner Johan Swärd presents some
of the around 50 grain varieties he maintains at a farm field day
Source: Biodynamic Association, Norway

does not prevent farmers from exchanging seeds among themselves. It is, however, illegal for farmers to sell seeds of protected varieties. This seems to be generally accepted among farmers in Norway as a legitimate way of ensuring breeders' rights.

In 2005, the Norwegian government decided to reject a proposed amendment to the law that would have brought about a significant expansion in the rights of plant breeders. Although Norway was a member of UPOV under the 1978 Act of the Convention (hereinafter called the 1978 Convention), the new law would have set the stage for Norwegian membership under the 1991 Act of the Convention (hereinafter called the 1991 Convention), which is far more rigorous (see p. 14). When the Norwegian government rejected the bill, one main argument was precisely the need to take Farmers' Rights into account.

Plant breeders' rights under UPOV

The International Union for the Protection of New Varieties of Plants (UPOV) was founded in 1961 to foster the development of new plant varieties and the trade in them by establishing a uniform system in member countries on plant breeders' rights to new plant varieties for a fixed period. In this way, the efforts of plant breeders could be recognised and compensated, and the system should stimulate further innovation in the field.[4] The patent system was not suitable here: the protection it afforded would be so strict as to deny plant breeders any opportunity to build on each other's work. An alternative system was needed. Under the 1961 Act of the UPOV Convention, plant breeders were granted broad exemptions from property rights. The same applied to farmers: they paid a licence fee when the seed was purchased after which they were free to use and exchange seeds as they wished.

The first Act of the UPOV Convention entered into force in 1968 and has been amended several times since then. Each time, the changes have reduced the options available to farmers. Most UPOV member states base their membership today on either the 1978 or 1991 Convention. After 1998, new countries may join only under the 1991 Convention. Norway was a member before that date, however, and retains the right to extend its UPOV membership under the terms of the 1978 Convention.

There are major differences between the 1978 and 1991 Acts of the Convention under UPOV. The most important in terms of Farmers' Rights is that farmers under the 1978 Convention could still save seed from their own harvest and sow it the following year, as farmers had done since the dawn of agriculture. Under the 1991 Convention, this is forbidden, although exceptions can be made for small-scale farmers if the seed is used on their own land to a limited extent and in compliance with plant breeders' interests. In practice, this means that in many countries a licence fee is imposed for this type of use as well. In Germany, for instance, farmers have to pay 80 per cent of the total cost of the licence if they re-use seed from the previous year's harvest for varieties of many species.[5] Another important difference is that under the terms of the 1978 version farmers may exchange seeds and propagation material from protected varieties but not under the 1991 UPOV Convention. Both Acts of the UPOV Convention forbid the sale of protected seed.

A proposal to strengthen plant breeders' rights in Norway

An amendment to Norwegian legislation was proposed in 2005 in response to the privatisation of the seed industry a few years previously. The government had expected the seed industry to adapt itself to market forces, allowing for a gradual reduction of government support. In the event, however, it proved impossible to recoup expenses through revenues from the licence fees sanctioned under the old law on plant breeders' rights. The seed industry therefore proposed that the law be amended, thereby enabling Norway to become a member of UPOV on the basis of the 1991 Convention. This would ensure necessary, albeit still insufficient, funding for plant breeding. The draft law was sent out for public hearing in January 2005. All the farmers' organisations and several voluntary organisations opposed the bill. Members of the scientific community warned against adopting it as well. The proposal was also debated in newspapers and on radio programmes. Two reasons in particular were cited for rejecting the proposal: first, the new law would limit the traditional rights of farmers to save, use and exchange seed and propogating material from their own harvest; second, the costs would be borne by the farmers since they would have to buy seed and propagating material every new season. For some plant species, small-scale farmers would be able to use seed and propagating material from their own harvest but only after paying a licence fee to the rights-holders.

The plant breeding industry supported the bill but proposed the addition of several more exemptions for farmers.

Draft law rejected on the grounds of Farmers' Rights

Following the parliamentary election in the autumn of 2005, which brought an alliance of social democrats, agrarian centrists and environmentalists (the Norwegian Labour Party, Centre Party and Socialist Left Party) into power, a former board member of the Norwegian Farmers' Union, Terje Riis-Johansen, was appointed Minister of Agriculture and Food. One of his first acts – in response to the discussion following from the public hearing – was to reject the bill on the grounds that it undermined Farmers' Rights. This was celebrated as a success among farmers in Norway.

Some months later, Riis-Johansen facilitated a transfer of funds to the breeding industry to compensate for loss of revenue. This was an important step and supported the breeding of varieties for which there is insufficient commercial demand but which are still needed in Norway. While Norway's plant breeding industry is small, it is nonetheless essential for agriculture in a country with such particular growing conditions. No matter how the law might have been changed, it would seem virtually impossible to recoup all the costs of plant breeding caused by more stringent plant breeders' rights. Therefore, plant breeding in Norway will have to continue to depend on government support. Rejecting the bill while ensuring sufficient funds for plant breeding: that was the Norwegian solution to balancing farmers' and breeders' rights.

Norway's approach to regulations on plant variety release and seed marketing

The rules for plant variety release and the marketing of seed and seed potatoes in Norway have changed radically over the past ten years. From a situation in which farmers could exchange all types of seed with each other and sell seed of non-protected varieties, almost everything was banned in 2004. Not only were farmers forbidden to sell seed among themselves – they could not exchange seed material or even give it away. Only government-authorised seed shops were allowed to sell seed but only seed of varieties on the official list of varieties. So stringent were the criteria for variety release that most of the older varieties, also popular older varieties from neighbouring countries, would be rejected. As they could also not be exchanged among farmers, farmers could continue to cultivate only what they already had on their farms in 2004. If they lost some varieties or lost interest in the work or for any other reason stopped saving and cultivating them on their farms, those varieties would go out of production and no other farms would be able to take over.

The reason for the new policy lay in Norway's membership of the European Economic Area (EEA); according to which Norway must implement all EU legislation that is declared EEA-relevant. Plant variety release and seed marketing are EEA-relevant issues, and thus all EU legislation on these topics has to be implemented in Norway as well.

Even if the new regulations were adopted in 2004, no effort was made to spread information about this. As Norway must keep amending its regulations continuously to implement EEA-relevant EU directives, it is difficult to keep track of all the changes. Not all amendments are sent out on public hearings and even if they are, the opinions collected have little real impact on the resulting legislation: the conditions are set in the EU. This is a serious democratic problem. The new regulations concerning variety release and seed marketing were not made subject to public hearings in Norway, and people were simply not aware of them initially.

Farmers gathered at the Organic Seed Days in Vestfold County on 25 January 2006, were taken by surprise when a representative of the Food Safety Authority informed them that it was now prohibited to exchange and sell seeds and that even offering seeds as a gift was forbidden. The farmers could simply not believe that this was true, given the Norwegian policy on Farmers' Rights and not least due to the recent success of rejecting the plant variety protection bill on the grounds of Farmers' Rights. The farmers kept asking and challenging the representative, stressing that this could not be real, it would be against Norwegian policy and that she had surely misunderstood. Finally, she herself began to doubt, and she promised to go back to the office and check. Nevertheless, her later answer left no doubts whatsoever: she had indeed been right, such was now the law.

These regulations would have made it almost impossible to conserve plant genetic diversity on farms, and it would have been only a matter of time before this work ceased altogether. Had it not been for the farmers' readiness to ignore the official rules, with the authorities indicating that they would turn a blind eye, work on crop genetic diversity on farms would have suffered a serious setback.

However, as we shall see, the Norwegian authorities wanted to find a solution, recognising that the regulations were clearly not in line with Norwegian policy. To understand the course of the events, and the final achievement, we return to where it all started.

Prohibition years 2004–2010

In 2003, the Norwegian Parliament passed the Food Act,[6] which replaced several earlier laws on agriculture, food production, food trade and food security. The Food Act also governs variety release and the marketing and production of seed and propagating material as specified in separate regulations. In this connection, the seed marketing regulations[7] were changed in 2004 to bring them in line with EU directives on seed marketing, in light of Norway's obligations as an EEA member.[8] And, because Norway unintentionally came to omit one EU definition's specification 'aimed at commercial exploitation', these regulations became even stricter. All forms of marketing of seeds in Norway were consequently banned, apart from marketing undertaken by government-authorised seed shops (Paragraph 4). 'Marketing' would now mean 'possession with a view to sale, offering for sale, distribution and the sale and any other form of transfer, with or without compensation' (Paragraph 3, Section n). In other words, farmers as well as hobby gardeners were prohibited from selling and exchanging seeds nor could they give seeds away.

It was even forbidden to save seeds with a view to selling, exchanging or giving them away.

The ban applied to virtually all kinds of food and fodder plants, whether protected by plant breeders' rights or not. The new regulations marked a radical break with customary practices in Norwegian agriculture.

In addition, two other important changes were imposed. The first change was that only seed of released varieties could be traded in Norway. A variety was 'released' if it was on the official Norwegian list of released varieties or on the EU's common lists of released varieties (Paragraph 15). To be approved, the variety needed to be distinct from other varieties, genetically uniform and stable in accordance with UPOV's technical guidelines and it had to pass the test of value for cultivation and use (VCU test) to determine whether it was to be deemed a 'useful' variety (most horticultural species and some other plants were exempted from the valuation regulation) in compliance with the regulation on plant variety release (Paragraph 3 and Paragraph 4).[9] The second change was that all seed should be certified in order to be eligible for trading, with the exception of berries, fruits, herbs and park plants (Paragraph 15 of the plant variety release regulation). 'Certified' seed is seed that has been bred under governmental supervision and classified in accordance with established quality standards (Paragraph 3, Section a). For the seed to be certified, it had to be shown to have stemmed from a released variety.

Most of the varieties Norway's biodiversity farmers have been working to conserve and develop do not comply with the criteria enumerated in these regulations. And so, the sale of these varieties was banned as of 2004.

The intention behind the seed marketing regulations was to ensure the highest possible standards of health and quality in the production and sale of seeds (Paragraph 1) – a matter of undoubted importance for all stakeholders in agriculture. Plant diseases and poor seed quality can have serious consequences for productivity, and legislation in this area has historically served to facilitate significantly higher production yields and improved quality of agricultural produce. Nonetheless, the new regulatory constraints represented a paradox. Plant health in the long run depends on sufficient genetic diversity from which to develop the varieties that have the necessary resistance to plant diseases and pests at any given time and are adapted to other environmental factors and needs. Eliminating the possibility of conserving and developing plant genetic diversity *in situ* on-farm means reducing the ability to ensure plant health in the future. These short-sighted regulations on plant health therefore clashed with society's longer-term need precisely to ensure plant health. For the first time, regulations intended to ensure plant health and seed quality effectively pulled the carpet from under their own goals.

In 2006, this situation was brought to the attention of the Ministry of Agriculture and Food.[10] The Ministry responded that it would have the regulations changed by the end of 2007 to facilitate the conservation of plant genetic resources.[11] The work on drafting the amendments was delegated to the Norwegian Food Safety Authority. However, it proved to take longer than anticipated, and the amended regulations were not in force until 2010 – due not least to negotiations in the EU on a new directive on conservation varieties.

Figure 3.2 Participant at a farm field day shows the height of the slash-and-burn rye
(svedjerug); an old variety that was re-discovered and became popular in
Norway
Source: Biodynamic Association, Norway

The EU Directive on Conservation Varieties

In 2008 the EU adopted a directive on conservation varieties that entered into force on 30 June 2009 (EU Commission Directive 2008/62/EC). As an EEA member, Norway is obliged to comply with this Directive. The Directive seeks to ensure the conservation and sustainable use of 'conservation varieties'. Such varieties may be cultivated and marketed even when they do not meet the general requirements for recognition of varieties and sale of seeds and propagating material (Article 2). Here the Directive sets out its own guidelines for the recognition and inclusion of 'conservation varieties' in national lists of varieties and the production and marketing of the seed material.

Negotiations preceding the adoption of the Directive were difficult because of competing interests. Key players in the European seed industry were pushing for a very tight regime: such liberal rules on conservation varieties would 'skew competition', they argued. Farmers' organisations and NGOs, on the other hand, wanted maximum legal manoeuvrability in the conservation and sustainable use of crop genetic diversity. Economic interests clashed with the public interest over conservation and sustainability, and here the former interests had a major impact.

These are the basic features of the EU Directive on Conservation Varieties:

- *Basic requirements:* Landraces and varieties, which are naturally adapted to local and regional conditions and are threatened by genetic erosion (Article 1) and are of interest for the conservation of plant genetic resources (Article 4.1) may be approved in accordance with the directive.
- *Compliance with DUS criteria:* Varieties must meet the normal requirements of distinctness, uniformity and stability ('DUS') for approval (Article 4.2). But in some cases (off-types), the uniformity requirement may be relaxed somewhat.
- *Genetic restrictions:* Approved varieties must be conserved in such a way as to ensure continued varietal identity and varietal purity and shall be inspected in accordance with given provisions to verify compliance (Article 19).
- *Geographical restrictions:* A 'conservation variety' shall be cultivated and marketed only in its region of origin, and seed may be produced only there (Article 11, Article 13). The region of origin must be identified prior to approval (Article 8) and may include more than one country.
- *Certification requirements:* The usual certification requirements apply here, with an exception as to the requirement on minimum varietal purity. Nevertheless, the Directive stresses that the seeds are to have sufficient varietal purity (Article 10).
- *Marketing requirements:* Seeds may be marketed only by authorised seed shops in the seeds' region of origin, with exceptions for cases in which a member state approves additional regions in its own territory for such marketing (Article 13). In other words, the prohibition on seed exchange between farmers remains in place under the new Directive. In order for farmers to be able to sell seed from their fields they must establish authorised seed shops.
- *Quantitative restrictions:* The quantity of seed marketed is not to exceed 0.5 per cent of the seed of the same species cultivated in the country in one growing

season, or alternatively a quantity necessary to sow 100 hectares – whichever is the greater quantity (Article 14). This is the general rule. For certain species (peas, wheat, barley, maize, potato, rape and sunflower) stricter provisions apply. In these cases, the percentage shall not exceed 0.3 per cent, or a quantity necessary to sow 100 hectares, whichever is the greater quantity. The total quantity of seed of conservation varieties marketed in each country shall not exceed 10 per cent of the seed used yearly of the species involved. If this leads to a quantity lower than that necessary to sow 100 hectares, the maximum amount of seed of the species concerned may be increased to the quantity needed for sowing 100 hectares.

Although these rules were designed to soften the previous regulations that were hindering the conservation and sustainable use of crop genetic diversity in agriculture, they are very restrictive nonetheless. First of all, seed exchange and marketing among farmers is still prohibited. Furthermore, since only approved conservation varieties may be marketed, all other varieties are left outside the scope of marketing and thereby also on-farm conservation and use in the long run. The requirements as to genetic uniformity remain quite rigorous, thereby pushing a large number of varieties of current or potential interest for farmers outside the market and outside the scope of further on-farm conservation and use. Since it is forbidden to market or use a conservation variety outside of its region(s) of origin, the traditional exchange of seed is considerably limited without any basis in plant health or seed quality concerns. In addition, there is a quantity restriction on the marketing of seed of conservation varieties, further limiting their use. And finally, the seed of these varieties cannot be marketed if the varieties have been further developed and thereby lost the properties on the basis of which they were originally approved.

The EU Directive on Conservation Varieties has been welcomed by plant breeding companies in Europe for regulating the seed market in such a way that all actors must follow the same rules. However, it has met heavy criticism from farmers' organisations, researchers and NGOs in the EU and in Norway for emplacing obstacles to the conservation and sustainable use of crop genetic resources, instead of promoting these as provided for in the Plant Treaty (see Andersen 2012; Winge 2012).

Relaxing the Norwegian regulations on variety release and seed marketing

After the regulations prohibiting farmers from following their customary ways of exchanging and using seeds became known in 2006, the NGO Oikos – Organic Norway took the lead in organising dialogues with the Food Safety Authority. As early as December 2006, Oikos sent a letter to the Ministry of Agriculture and Food expressing concern over the regulation on plant variety release and seed marketing. The problems were featured in several articles in Oikos journals. In autumn 2008, Oikos organised two meetings with farmers' organisations, other stakeholders and researchers to discuss how to proceed, with the Norwegian Food Safety Authority represented at one of these meetings. Oikos continued its commitment, hosting and attending additional meetings, and maintaining contact with the Norwegian Food

Safety Authority. These meetings and contacts helped to facilitate a shared under-standing of the challenges that the regulations represented for biodiversity farming in Norway.

In January 2009, a one-day meeting convened at the Fridtjof Nansen Institute near Oslo brought together a total of 60 participants on the topic of Farmers' Rights in Norway. All relevant actors were represented including the political leadership of the Norwegian Ministry of Agriculture and Food, representatives of the Food Safety Authority, farmers' organisations, the breeding industry and seed companies. This provided a unique opportunity to discuss and deliberate together on the issues involved and helped to enhance a shared understanding of the challenges involved in regulating the seed market.

From time to time there has been media attention as well, and several articles in newspapers and magazines have helped to draw attention to the topic as well as the process of amending the relevant regulations.

The dialoguing among stakeholders during the processes of amending the regulations on variety release and seed marketing in Norway proved constructive and positive, fostering a shared understanding of the needs and challenges. Nevertheless, the great question remained: how could the authorities manage to combine consideration for Norwegian opinions with the country's obligations under the EEA Agreement? These concerns were not lessened after the adoption of the EU Directive on Conservation Varieties, as there was great scepticism of the Directive among central stakeholders in Norway.

On various occasions, central biodiversity farmers maintained that there were serious problems with the Conservation Varieties Directive. First, they feared that many of the crop varieties of relevance to them would not be approved because they were too genetically heterogeneous. Further, they stressed that Norway had lost most of the crop genetic diversity that had been present before the advent of professional plant breeding. In order to develop crop genetic diversity adapted to the varying growing conditions of this long-stretching high-latitude country, it was important to have access to older varieties from neighbouring countries and other countries with similar growing conditions. Such varieties could also have genetic properties from Norway, due to the seed exchange that had taken place across borders over the centuries. With the new EU Directive on Conservation Varieties, however, all this would become impossible due to the provision that only the region or regions of origin would be allowed to market and grow the respective varieties. Another serious concern for the farmers was the fact that they were not entitled to further develop the varieties for marketing, as these were now required to be conserved in the way they were when approved. For most biodiversity farmers that meant that the whole motivation for growing diversity was abolished. What they wanted was to work on the varieties through selection; further adapting them to differing environmental conditions and consumer demands in order to improve them and thereby contributing to the further development of crop genetic diversity in Norway. These concerns were acknowledged by the Food Safety Authority even though accommodating them was difficult because of Norway's obligations under the EEA Agreement.

In September 2009, the Food Safety Authority sent the draft amendments 'on the approval of conservation varieties and production and sale of seed and seed potatoes of those varieties, and on exceptions for non-commercial trading of seed' to a public hearing.[12] A total of 20 organisations and bodies responded with comments on the draft. Of these, six were positive and/or suggested further improvements on technical issues. The other 14 were essentially in favour of the proposal but were critical to several points. Comments and suggestions were particularly concerned with restrictions relating to region of origin, uniformity, value for conservation, marketing quantities and specific restrictions related to seed potatoes, as well as the need to develop varieties in farmers' fields and be compensated for the costs associated with establishing authorised seed shops and the approval of varieties.

Although the hearing bodies put in a great deal of time and effort examining these issues and offering constructive suggestions, hardly any of their ideas were taken into account in the end. The few changes that were made by the Norwegian Food Safety Authority as a result of the public hearing were limited to improving the wording and providing clarification but stopped short of substantive changes to the proposal. EU regulations are to have priority over proposals from Norwegian hearing bodies – even if the latter are unanimous – in case of disparities between a Norwegian proposal and an EU regulation.

On 30 April 2010, Norway's Ministry of Agriculture and Food endorsed the changes to the Regulation of 13 September 1999, No. 1052 on Seed Marketing; to the Regulation of 2 July 1996, No. 1447 on Seed Potatoes; and to the Regulation of 1 October 1999 on the Testing and Approval of Plant Varieties, all in accordance with the recommendations of the Norwegian Food Safety Authority. This action served to soften the prohibitions of 2004; and as we will see, the Norwegian Food Safety Authority went to great lengths to accommodate EU regulations on the principles of sustainable management of crop genetic diversity, even if Norwegian stakeholders had suggested more radical measures.

More EU seed directives to be implemented: vegetables and fodder-plant mixtures

As early as November 2009, the EU passed Commission Directive 2009/145/EC on traditional vegetable varieties. The purpose of this directive was to facilitate the sale of seed with no intrinsic value to commercial producers but which had been adapted to thrive in particular conditions related to climate, soil or cultivation. In Norway, these varieties are called 'traditional vegetable varieties'. Directive 2009/145/EC is more liberal than Directive 2008/62/EC on conservation varieties, as there are no requirements relating to region of origin or threat of genetic erosion as criteria for inclusion on the official list of varieties.

There are, however, other drawbacks. The criteria for including vegetable varieties on the list of varieties as 'traditional' varieties, thus allowing for legal marketing, are that they should have no intrinsic commercial value and may be sold only in small packets of 5, 25 and 250 grams (depending on the species). The seeds of traditional varieties can be sold to gardeners on a hobby basis, but not to farmers for commercial use in agriculture. These two provisions (small seed packets and selling only for non-commercial use) are intended to limit the dispersal of these varieties. Importantly, the regulations on traditional vegetable varieties will not help farmers to preserve crop genetic diversity in vegetables, as farmers are debarred per definition from using seed of such varieties.

Mention should also be made of EU Commission Directive 2010/60/EU of 30 August 2010 providing for certain derogations for marketing of fodder plant and seed mixtures intended for use in the preservation of the natural environment. It is written along the same lines as the EU Directive on Conservation Varieties and is subject to implementation in Norway as well.

Current rules and practice to date

The new regulations on conservation varieties and traditional vegetable varieties are enacted in three regulations in Norway:

- Variety Release Regulation concerning approval of plant varieties for listing on the Norwegian official list of varieties;
- Seed Marketing Regulation concerning the marketing of seed;
- Seed Potato Regulation concerning the marketing of seed potatoes.

The variety release regulation applies to all species covered by the seed marketing regulation and seed potato regulation. The seed marketing regulation covers the most important named species of cereals, fodder plants and grass for green spaces, beets, oleaginous and fibrous plants and all kinds of vegetables and berries, fruits, landscape plants, spice plants, medicinal plants and ornamentals. The seed potato regulation covers all seed potatoes. As a result, only a very few plant species are not covered by the regulations. One example is emmer (*Triticum dicoccum*), in the wheat family. Whereas ordinary wheat (*Triticum aestivum*) is covered in the regulations, emmer is not. The same applies to einkorn wheat (*Triticum monococcum*). Both species are grown on several farms in Norway. Otherwise, however, the vast majority of food and forage plants cultivated by biodiversity farmers are currently covered by the regulations.

As of 2010, Norway's farmers were again allowed to save, exchange and sell seed, with the exception of seed potatoes, on a non-commercial basis. This is specified in the legislative pre-works to the 2009 amendments of the said regulations, but exactly what 'non-commercial' is supposed to mean is not defined. There may be good reasons for not establishing an interpretation here since this remains something of a 'hot potato' in the EU, and the Norwegian Food Safety Authority has obviously wanted to give Norwegian farmers as much flexibility as possible in this area. Non-commercial trading of seed of old varieties (exchange and selling in small quantities) among biodiversity farmers appears to be on the increase.

Farmers may also establish authorised seed shops for conservation varieties on relatively easy terms, such as their being registered in Norway and led by a manager with appropriate qualifications. This does not apply to seed potatoes although the regulations do allow the Norwegian Food Safety Authority to let gene banks and the like sell seed potatoes under certain conditions. The company Økologisk Spesialkorn

AS (Organic Special Grain), established by a group of biodiversity farmers to the north of Oslo, was the first to be authorised as a seed shop for conservation varieties in 2011.

According to the Norwegian regulations, the sale of seed of conservation varieties from authorised seed shops is conditional on the release of the varieties as conservation varieties. This is in turn dependent on the consideration of value for conservation and the identifiable area of origin within Norway, and the variety must be maintained in this area. Performing a VCU test is not required, provided certain other conditions regarding documentation of utility are met. The variety must not have been protected by plant breeders' rights or included on the official Norwegian list of crop varieties in the two years preceding application.

An applicant seeking recognition for a conservation variety must provide information on the area of origin; name and any known synonyms or local names; description of the variety utility value; growth characteristics and conservation value; as well as the planned approach to preserving the variety. Assurances must be given that the characteristics will not change but be maintained in the variety's approved form. Recognition currently costs approximately US$100 per variety.

NordGen has offered to act as applicant and conservationist of the seed-propagated varieties and may also assist others who wish to apply for recognition of such varieties in Norway. The Norwegian Genetic Resource Centre offers the same services in respect of potato varieties.

As of December 2012, seven conservation varieties had been approved and included on the official Norwegian list of plant varieties: four wheat varieties, one barley variety and two potato varieties. About 40 varieties of spring wheat, oats, barley, peas and beans were on file with NordGen awaiting description in preparation for an application for approval.

The potato gene bank under the Norwegian Genetic Resource Centre will have around 90 potato varieties by the end of the approval procedure for the current varieties. All varieties are stored *in vitro* at Bioforsk (Norwegian Institute for Agricultural and Environmental Research), and the Overhalla Cloning Centre will produce clones for distribution to ensure access to virus-free seed potatoes of approved varieties. Several options are being considered to facilitate such access. The Norwegian Genetic Resource Centre has been spearheading this work.

Vegetable varieties may be registered either as conservation varieties or traditional vegetable varieties. As of 2012, no varieties of vegetables had been registered.

The Norwegian Food Safety Authority is currently in the process of approving varieties. This can be a daunting task at times because of documentation requirements and challenges related to assessing the documentation in accordance with the regulations. In this sense, the rules are demanding for the authorities as well.

Concluding remarks

Norwegian authorities have gone further than their counterparts in most other countries in Europe in accommodating Farmers' Rights to save, use, exchange and sell seed and propagating material. Farmers in Norway are still allowed to save seed of varieties protected by plant breeders' rights, and they may use the seed in the following season and exchange it among themselves. By contrast, in most other countries in Europe, farmers may not exchange seed materials among themselves, and saving seed of such varieties and using it in the following season is prohibited or requires a licence.

As regards plant variety release and seed marketing regulations, the Norwegian authorities have also gone further than most other governments in Europe. Farmers in Norway are allowed to exchange and sell seeds and propagating material (except seed potatoes) on a non-commercial basis among themselves.

It has become easier for farmers to set up seed shops for conservation varieties. Such varieties need to be registered in the official Norwegian list; for several varieties, the seed also needs certification. The requirements are less stringent for conservation varieties, but the Norwegian Food Safety Authority has signalled that it intends to exercise considerable flexibility here. It is still too early to tell what impact this will have as only a few conservation varieties have been registered so far. The Norwegian Food Safety Authority is likely to go as far as they can to allow approval of as many varieties as possible, and NordGen and the Norwegian Genetic Resource Centre will help in preparing variety descriptions and other information likely to increase the chances of formal approval.

Time will tell what the approval level will be, the bureaucratic costs involved and whether varieties are approved quickly enough in relation to demand and the interests of farmers. This is important to prevent farmers from sustaining economic losses and to enable them to promote diversity at their preferred pace. Time will also tell whether quantity restrictions will be applied and whether regional restrictions on cultivation and seed production will have any effect. And finally, only time can reveal the impact of the requirement to save varieties in the form in which they were registered. Will it be possible to comply with these requirements when applied to varieties with relatively large genetic heterogeneity and varieties which are still evolving naturally? Will the rules be enforced? Will the rules make it less attractive for farmers to promote crop genetic diversity, when they stop growers from contributing to the development of genetic diversity and adapting varieties to current needs and, not least, to shifting environmental conditions?

It is too early to say anything about the consequences as yet. Instead, we have indicated factors that could impinge on Farmers' Rights to save, use, exchange and sell seed, thus limiting their ability to save, develop and sustainably use genetic diversity in the future. There are many warning signs in the regulations currently in force in Norway. If these warnings prove true, it would significantly constrain the opportunity to preserve crop genetic diversity, use it and offer it to users. Thus far, the Norwegian Food Safety Authority has done its utmost to prevent this scenario from materialising. It is because of this relative success that this example from Norway can be classified as a success story.

Lessons for other countries

An important question: what lessons may this story provide for other countries? Norway's experience with regulations on variety release and seed marketing highlights the great challenges that the current trend of seed regulation poses to the further conservation and sustainable use of crop genetic resources in line with the Plant Treaty. It has also shown how important stakeholder consultations and dialogue are, in ensuring regulations that seek to accommodate all concerns. We have seen that it is indeed possible to create legal space for farmers to continue their contributions to the conservation and development of crop genetic diversity, even under such difficult framework conditions. If more European countries would follow suit, this could establish a practice that would contribute to enhancing the legal space for Farmers' Rights to save, use, exchange and sell farm-saved seed. All the same, an important question is whether this is the right path to follow, in terms of the objectives of the Plant Treaty. Perhaps other solutions should be sought to regulate the seed market? The EU has begun a revision process of its seed legislation, which may provide new answers to this.

What can be learned from the Norwegian experience regarding the plant variety protection legislation? Norway today is an affluent country and can afford to support its breeding industry while at the same time ensuring that farmers' and breeders' rights are balanced in a way conducive to the conservation and sustainable use of crop genetic resources. Other countries do not have similar financial muscle – so what about them? This gives rise to further questions: what should be the division of labour between the state and the breeding industry? And whose responsibility is it that there is sufficient diversity of varieties adapted to the environmental conditions and consumer demand? Should it be left to the 'invisible hand' of the market – or does the state have a role to play? What would happen if the market forces alone should rule the game?

In the case of Norway, that would drastically reduce the number of varieties available to farmers. Therefore the Norwegian government supports the breeding of varieties that are not commercially viable but for which there is still important demand. Perhaps this is a possibility that more countries should consider, as it is a matter of high importance for agricultural production and access to plant genetic diversity now and in the future.

Only a few decades back, plant breeding was a public responsibility. Now the pendulum has swung to the opposite extreme – to privatisation of the plant breeding industry. It may be time to consider a middle course where both have a role to play but where the state has a particular responsibility to ensure – through regulations and economic incentives – that sufficient crop genetic diversity remains available, while at the same time balancing Farmers' Rights and breeders' rights.

Notes

1   This chapter is based on a report on plant genetic diversity in agriculture and farmers' rights in Norway (Andersen, 2012), which is the product of a long and highly participatory process. In addition to document analyses, central inputs have come from participation at meetings, seminars and consultations with stakeholders as well as interviews with farmers, representatives of farmers' organisations, the seed industry and relevant authorities. The one- day meeting convened at the Fridtjof Nansen Institute in January 2009 (presented later in the text) provided a unique opportunity for discussing and sharing viewpoints on the central issues involved in this present chapter.

2   'Biodiversity farmers' refers here to farmers who actively contribute to the conservation of genetic diversity in agriculture by cultivating older plant varieties, conservation varieties or other varieties not included on the official list of varieties and/or by maintaining biodiversity-rich meadows and pasturelands through traditional methods. The term may also refer to farmers who keep older breeds of livestock.

3   Lov om Planteforedlerrett (LOV 1993-03-12 nr. 32) www.lovdata.no/all/hl-19930312-032.html

4   This and the following section are based on Andersen, 2008, pp. 146–160.

5   The German law on the protection of plant varieties allows farmers to save seed from protected varieties if they pay a fee to the rights-holder. The Federal Court of Justice (Bundesgerichtshof) has set the fee at 80 per cent of the full licence. Sources: Bundesministerium der Justiz: Sortenschutzgesetz. Adopted 1985, latest revision 2008 (www.bundesrecht.juris.de/sortschg_1985/BJNR021700985.html) and Deutscher Bauernverband (2007): Bundesgerichtshof: Pauschale Nachbaugebühr in Höhe von 80 Prozent zu hoch – DBV sieht Vorteilhaftigkeit der Rahmenregelung bestätigt (www.pressrelations.de/new/standard/result_main.cfm?pfach=1&n_firmanr_=100982&sector=pm&detail=1&r=287153&sid=&aktion=jour_pm&quelle=0).

6   Lov om matproduksjon og mattrygghet (LOV 2003-12-19 nr. 124).

7   Forskrift om såvarer (FOR 1999-09-13 nr. 1052).

8   While agriculture is not included in the EEA agreement, plant varieties and seed are, so the EU rules relating to seed material apply in principle in Norway as well. Forskrift om såvarer (FOR 1999-09-13 nr. 1052) makes reference to EU rules in this area.

9   FOR 1999-10-01 nr. 1069: Forskrift om prøving og godkjenning av plantesorter.

10   Letter from the author to Secretary Per Harald Grue, Ministry of Agriculture and Food, 1 December 2006, inquiring into the interpretation of the seed regulations and their harmonisation with the Plant Treaty.

11   Letter from Head of Section Kjell Nyhus and Marianne Smith, Ministry of Agriculture and Food to the author, 9 February 2007 (LMD reference: 200602968/MSM)

12   While recognition of plant varieties and the sale of seed are covered under the EEA Agreement, seed potatoes are not. Notwithstanding, the Seed Potato Regulations were included in this proposal in order to ensure consistency across the relevant Norwegian regulations.

4   Conservation of local varieties in the Basque Country in spite of legal restrictions

Tone Winge and Helen Groome

In the Basque Country of Northern Spain, local varieties are being conserved through documentation and use, despite detrimental laws and a lack of legal space for the traditional practice of seed-sharing. Spain's seed law (Law 30/2006 on seeds, nursery plants and phytogenetic resources) prohibits the exchange and sale of seed from unregistered varieties; and since most local varieties are not registered and are often too heterogeneous for registration under the current system, it is illegal to sell and exchange seed from these varieties. Because exchanging, selling and buying seed is essential to the continued cultivation of local varieties, thus also conserving them for future generations, a local organisation called the Basque Seed Network has been documenting varieties, distributing seed and carrying out awareness-raising activities.

Maintaining the local heritage of plant varieties is seen as very important by the members of this relatively small but active network. Their activities are geared towards finding and documenting what is left of this diversity and making sure that it will not be lost. Their main approach involves conserving the local varieties through use and cultivation, and to achieve this the distribution and exchange of seed from unregistered varieties is central. Therefore, the network also engages in dissemination of information and lobbying, to enable them to carry on their work despite the current legal situation, and in the hope of making conservation easier by getting the legislation changed.

This approach to the realisation of Farmers' Rights is useful when there is limited legal space for the seed practices necessary to sustain local varieties. Protecting and promoting the rights of farmers to use, exchange and sell farm-saved seed is widely regarded as one of the most central aspects of realising Farmers' Rights, but in many parts of the world this is becoming increasingly difficult due to seed laws and plant-variety protection laws. The following story can offer some useful lessons for civil society organisations looking for ways to deal with this problem under similar circumstances.[1]

The Basque Seed Network

In 1996 the Basque Farmers' Union (EHNE) organised an agrobiodiversity meeting. At this meeting, several interested individuals decided to work together on questions of plant diversity within food production. It was decided that the core activities of the group would be to find out what still existed of plant genetic diversity through seed surveys, to disseminate information through a seed-savers' manual and to raise awareness through talks and workshops. Thus the Basque Seed Network was formed.

Until 2001 the group was organised under the umbrella of EHNE, which provided legal protection and a formal identity – through the union they could, for example, apply for funding. Some members of the network represented organisations working within related areas, while others had no specific affiliation. Then in 2001 the network became independent of the union and was set up as a separate association. Since then the network's activity portfolio has expanded to include other subject areas, among them the issue of legal space and how the network can operate under the current legislation on the marketing of seed and plant propagating material.

The first seed survey organised by the network to get a picture of existing agricultural diversity was quite small-scale, but since then more comprehensive surveys have been carried out with funding from the Environmental Department of the Basque government. For this work and its other activities, the network relies on motivated volunteers – which also makes its conservation work quite vulnerable. The network has assembled collections of local varieties that it maintains, and in 2008 it merged the various collections held by the different organisations represented in the network. This collection is now being taken care of by Neiker, a publicly financed research institution and collaborating partner of the network.

As seen by Marcelino Santiago Mimbres, one of the network's members, a very important mission of the network is to coordinate different interests and organisations, seeking to balance the influence of the agribusiness industry and their philosophy. He feels that the big companies and the government are imposing the view that modern varieties are better than traditional varieties – and here he does not agree. He finds the local varieties superior in terms of taste and nutritional value and thinks that it is important to underline that these varieties are directly useful, and not conservation-worthy simply because they happen to be local. Indeed, many of the members agree on the superiority of local varieties compared to imported ones. This, the members feel, is related to local adaptation – local varieties need less chemical attention because they are better adapted to the environment – as well as taste.

The current goal of the network is to maintain local varieties and ensure consumer interest in the end products so that maintenance can become self-sustaining. In addition, its members see themselves as involved in a wider effort to change attitudes to life – including food and how food is produced – and, in particular, to change what they see as the wasteful and polluting habits of food production and consumption that characterise industrialised societies. The Seed Network focuses on the link between local crop varieties and the social quality of food. Network members feel it is important to push ahead rather than wait for the government to 'catch up': action is urgently needed now, before the plant genetic heritage is lost.

The individual members of the Basque Seed Network became engaged in the issue of agrobiodiversity and involved in the network for various reasons. Some feel that farmers are losing control over the agricultural development and the varieties that are being grown and are concerned about the situation for local varieties. The difference between animal and plant biodiversity has also been noted by some members, and these members would like to see the same attention bestowed on plant genetic resources and local plant varieties as on local animal breeds. Others have slightly more political reasons for their involvement in the network; for instance, they may be opposed to the power of agribusiness and multinational companies and find it important to maintain the local heritage in the face of globalisation. Local varieties of plants are then seen as part of that local heritage which also includes language, culture, etc. The work of other organisations has also partly served as inspiration for the network and its members. One member described how meeting with Henry Doubleday Research Association (HDRA) staff in the 1990s made her think about the conservation of plant genetic resources. The Basque Seed Network now works with various organisations nationally and internationally, and it coordinates with other seed networks at the national level.

Since its establishment the organisation and composition of the network have changed. Some previous members, like two representatives from an environmental group, are no longer involved, while new members have joined. The extent of cooperation and coordination with other groups and organisations also working on conservation of the local plant genetic diversity has increased since the network was first formed. One factor that contributed to this was an evaluation the network had to do in connection with receiving a grant from the Environmental Department of the Basque government. Through this evaluation, the various groups got together – and the result was greater interaction. Attending and organising fairs has also affected the organisation of the network. Attending the annual plant fair held in Markina in the Bizkaia province has made their work more focused, whereas organising a national biodiversity fair in 2003 (a fair organised each year in Spain, in different locations) proved to be a challenge to their organisational set-up. The big budget and heavy workload put considerable pressure on the network and its members, and since then activities have mostly been on a smaller scale.

One of the active members, Aurelio Robles Loma, underlines the serious and methodological approach of the network, which he feels is recognised by their collaborating partners as well. He also views the network as a horizontal organisation, and considers this is important because it allows every opinion to be heard.

The network has about 90 members, with a core of five or six very active members. Around half belong to organisations like EHNE, various organic farmers' associations or environmentalist groups, but no one who wishes to join the Seed Network has to disclose such information, so the exact figures are somewhat uncertain.

Legislation and conservation

As a member country of the European Union, Spain is required to implement EU directives domestically. This means that in Spain, as in the rest of the EU, there are strict limitations regarding the production and sale of seeds. In Spain the seed law (30/2006 on seeds, nursery plants and phytogenetic resources) has been interpreted to mean that neither sale nor exchange of seeds from unregistered varieties is allowed. According to Bernardo Samaniego Gil, who works at the seed division of the Department of Agriculture of the Basque government, seed exchange and sale is defined as a commercial activity so the exchange of unregistered seed is therefore illegal.

With Commission Directive 2008/62/EC the EU has been moving towards greater understanding regarding the role of what it terms 'conservation varieties'. Conservation of genetic diversity has been the backdrop for this development. In Commission Directive 2008/62/EC, conservation varieties are defined as 'landraces and varieties which are naturally adapted to local and regional conditions and threatened by genetic erosion'. The Directive provides certain derogations for acceptance in the national catalogues of such landraces and varieties and the marketing of seed and seed potatoes of such landraces and varieties.

However, conservation varieties must also meet quite strict criteria as to variety release and certification. In addition, marketing and use of the varieties classified as conservation varieties is limited to the areas defined as the regions of origin for the variety in question, and only limited quantities of seed may be used. Because of the way conservation varieties are defined in the Directive and the limitations emplaced, it is not permissible for farmers to develop these varieties further – and that can be seen as a serious impediment to the continuous adaptation of crop genetic diversity to changing environmental circumstances.

Taken together, all these limitations mean that even after the EU Commission Directive on conservation varieties has been implemented in Spain it will be difficult for those involved in the Basque Seed Network to maintain the local varieties. Mr Samaniego Gil of the Basque government's Department of Agriculture does not think that the new directive will do much to change the situation with regard to the marketing of seed and that although it might make it easier to register local varieties and it does introduce the concept of 'conservation varieties', very little will be different as a result.

In Mr Samaniego Gil's opinion, seed policies to date have focused on protecting the buyer and producer and ensuring the quality of the seed. Although finding the work of the Basque Seed Network interesting, he sees it as being in conflict with 40 years of practical developments. He feels that more debate is needed about the issue of seed exchange and that this should be a discussion that takes into account the needs of farmers and consumers to a greater extent than today. The way forward for the network, as he sees it, would be to change the law – but since the regulations originate at the EU level that is also where the changes will have to be made.

However, Spain has other international obligations as well, and as a contracting party to the Plant Treaty (adopted in 2001) the country has committed itself to the realisation of Farmers' Rights and to ensuring the sustainable use of its plant genetic resources. It therefore has to find a balance between its commitments to the EU and to the Plant Treaty. This should be achievable, especially since the EU itself also is a contracting party and obliged to uphold the provisions of the Plant Treaty. At present though, since the Commission Directive on conservation varieties has failed to address sufficiently the problems that current seed legislation creates for the sustainable use of plant genetic resources and Farmers' Rights, it is up to individual countries like Spain to find ways of balancing their various commitments.

Despite the legal situation there are still people, also outside the seed network, who maintain local varieties by saving and sharing seeds. Most of them are middle-aged or elderly farmers and gardeners, and many fear that the varieties they cultivate will become lost when they themselves can no longer continue their work. Their rationale for cultivating these varieties and practising seed-saving and exchange varies. For Manuela Palenque, who lives in Karrantza in the Bizkaia province, usefulness is the main criterion when it comes to the local varieties she maintains. She grows a range of vegetable varieties for home consumption and has a pragmatic attitude. If she did not find these traditional varieties useful, she explains, she would not bother cultivating them. In addition, the cost-saving aspect is important. Because Ms Palenque saves seed from her own harvest, she does not have to spend money on seed, and this is one of her main motivations for saving and exchanging seed.

Although local traditions are perhaps not as vibrant as before, and many farmers have given up saving seed and cultivating local varieties, the existence of a strong tradition regarding the maintenance of traditional varieties in the Basque Country is still evident. Even today, you can stroll into a random farmers' market and find people who grow local varieties. At the farmers' section of the local market in Azpeitia in the Gipuzkoa province in late December 2008, for instance, it was quite easy to find a handful of farmers who cultivated many local vegetable varieties.[2] Sometimes these local varieties are cultivated in combination with modern varieties; whereas other farmers focus solely on traditional varieties. These farmers and gardeners who maintain their own local varieties usually save seed for their own use, although they do not necessarily exchange seed with others to a great extent. Sometimes farmers and gardeners cultivating local varieties buy plants or seed from others who also maintain local varieties, especially if they do not have the time to save their own seed. Time is often a constraining factor.

The degree to which the customers of these farmers realise that they are buying local varieties differs. Some farmers have regular customers who know that these vegetables are local and who might have a particular interest in local produce and choose to buy the vegetables precisely because these are traditional varieties from the area, but this is not always the case.

Selling produce can also be a way of distributing local, unregistered varieties without breaking the law. Interested buyers can then save seed from the produce they have bought and continue to cultivate these varieties. However, this does not work for all crops – only those for which viable seeds are present in the fruit or vegetable that is bought. Because of the legal restrictions and the lack of interest among members of the younger generations, many varieties are in danger of being lost when the ageing farmers and gardeners who have been maintaining them are no longer able to do so.

One thing most of the farmers and gardeners cultivating traditional varieties have in common is that they are generally not aware of the illegal aspects of exchanging and selling seed from unregistered varieties. Moreover, there is little sympathy for these rules among the farmers in question when they do find out about them. When told about the seed legislation and how it outlaws sale and exchange of seed from unregistered varieties, Ms Palenque's reaction was that these rules were strange and unnatural. Even when she buys seeds and vegetables at the local market it is from unregistered varieties. Seed saving and exchange has always been practised she argues, and she cannot see why this should not be allowed.

Activities of the Basque Seed Network

The activities of the seed network focus mainly, but not solely, on *in situ* conservation and have moved through different phases. Documenting the still-existing local varieties through the earlier-mentioned surveys is central. In connection with these surveys, seed samples of the documented varieties have been collected, and as noted, a collection of these is being taken care of by Neiker, a publicly financed research institution. Neiker is in charge of the material and distributes it upon request. Responsible for the gene bank and catalogue that the seed network has deposited at Neiker is Mr Jose Ignacio Ruiz de Galarreta. As long as there is a surplus, he sends out propagating material to people who make requests through the network. However, only one part of the 'variety collection' which the network has mapped through their surveys is being kept at Neiker. Other seeds are kept by traditional users and new 'seed guardians' *in situ*, and one or two members also keep small *ex situ* seed banks. However, the idea is to try to merge all *ex situ* collections.

Mr de Galarreta considers the network to be quite successful because of what it has achieved in terms of *in situ* conservation. As a result of the efforts of the network, local varieties are still being grown by farmers. He also sees the surveys and their identification of local varieties as successful. Although the success has been on a modest scale, he underlines that what distinguishes the gene bank that he maintains on behalf of the Basque Seed Network from other gene banks is that all the varieties are being cultivated, something he feels is both a part of and a reason for its success. Despite constraining factors such as a lack of resources and people, the network manages to get the propagating material out to farmers.

Because the varieties being distributed are unregistered, traditional varieties, there are legal aspects that could mean problems for this work. Maintaining the gene bank is legal, but the distribution of propagating material under the current law is another matter. So far, this is not an issue that Neiker and Mr de Galarreta have considered to any extent. He thinks that because the varieties are local they are free and distributing them should be unproblematic. However, as long as these varieties are unregistered, their distribution is in fact illegal under the seed law – unrelated to the fact that they are not covered by plant breeders' rights. This illustrates some of the problems the seed network and similar organisations may encounter due to the type of legislation prevalent in Europe.

As part of their conservation work, the network also has an urban allotment project that functions as a living gene bank for some of the local varieties mapped by the network. In Vitoria-Gasteiz, the town council runs an urban allotment project where the plots are distributed for free to individuals and groups. The Basque Seed Network is among the groups to have been given such a plot. On part of their plot the network is building a greenhouse. The intention, in addition to having the plot function as a living gene bank, is to use the network's presence to raise community awareness more generally. When the network has a surplus, it gives away seed from the crops it grows there to those who run the allotment project on behalf of the environmental department of the town council.

The allotment project organisers also give seeds from unregistered varieties to students that attend their classes and to other plot holders. According to the seed network the legality of this practice is not really seen as an issue, given that seeds have been exchanged for centuries; therefore, ethical considerations often outweigh legal considerations in many peoples' minds.

The seed manual first published by the network in 1999 has had a greater impact than originally expected. It has been quite popular and many farmers who carry out or are interested in seed saving say they have found it very useful. The inspiration for the manual came from a similar seed manual published by the HDRA, which in turn had been based on a publication by the Australian Seed Savers Network. Some of the farmers now associated with the network got to know about its work through this seed manual and one of them, Miguel Arribas Miranda, even found out about the network due to the seed manual while he was living in the Madrid area. He then contacted the network because he was interested in receiving seed – and that was the first time the network came in contact with farmers it had not known already. From the seed manual he had got the impression that the network was more consolidated than it actually was, but being in contact with them provided him with a more realistic view of how it operates.

A more recent activity of the Basque Seed Network is the orchard project that was started in 2006. As part of this project, old fruit-tree varieties are being conserved through documentation, restoration and distribution. The project area is located in a natural conservation park, Parque Natural de Valderejo; as of late 2008 it covered about half of the 3,500 hectares that make up the park. The area in question was made a national park back in 1992, but there were no conservation activities until the orchard project was launched. Most of the population left the area in the 1950s and the fruit trees remained basically untended. In addition to recovering local fruit-tree varieties, the project also aims to spread the varieties among the local farmers so these varieties can be used to generate income and to promote local development for the remaining people of the surrounding area. Thus, the orchard project is partly a rural development project and partly a conservation project.

The project was initiated by the local government because the national park management plan stipulates that the wild and domesticated plant life of the park should be surveyed and studied. Not knowing who might be able to do this type of work, the local government got in touch with Marcelino Santiago Mimbres from the seed network and together with Aurelio Robles Loma he undertook a small inspection to get a first impression of the park's remaining diversity. This inspection showed that the fruit trees in the park were of great interest, so it was proposed that a study should be done of the whole park. The park contains a range of climatic zones and is therefore rich in biodiversity, even though this area has not been historically important for food production. The study has been financed through the social fund of a bank via the local provincial government.

Figure 4.1 Apple varieties from the Basque Country
Source: Helen Groome

Mr Loma went on to join the seed network and became responsible for the orchard project.[10] The project surveys are carried out in a rather simple manner: the project leaders go around the park looking for trees and talk to the people in the area about their properties. Once a tree has been identified, it is reproduced and analysed. In order to maintain the genetic make-up of the trees, vegetative, asexual reproduction is used. Pear, apple, plum and gooseberry varieties are among the fruit trees to be documented by the project, and many of the re-discovered local varieties have already proven to be useful. When identified the local varieties are multiplied and made available. The orchard project now has a nursery containing about 600 plants.

To make the varieties available to the public in general will probably require permission from the authorities. Mr Loma does not want the collection to become a museum: it is important that the varieties are used. Among other things he wants to get cooks interested in these re-discovered varieties. A fruit catalogue has been made to present the varieties that have been discovered so far. Many of these varieties are quite resistant/tolerant to diseases and pests, and should therefore prove useful to fruit growers in the area.

Alongside the botanical surveys of the fruit trees in the park, an ethno-botanical inspection has been carried out to map the associated knowledge. Through this process, a substantial amount of information has been collected about the uses of the fruit and many interesting practices have been discovered, including techniques for fruit conservation. In conducting the ethno-botanical survey Mr Loma and his project team talk mostly to elderly people, as they are the ones who possess what remains of knowledge about the fruit trees. The fruit trees depend on them for their survival; the inspection made it clear that when these elderly people are no longer there, the trees will be in danger as well. What is being done in the park because of the orchard project is therefore a matter of urgency. Although much ground has been covered since the project was started, Mr Loma is still worried about trees dying before the project can get to them.

Apart from the seed network and the provincial government, the Public University of Navarre is also involved in the project. When the project was first initiated it was felt that some scientific guidance was needed and the Basque Public University was approached, but this university did not have the necessary resources and was therefore not interested. However, the Public University of Navarre was and its Department of Crop Production (Section Wine and Fruit) is now responsible for the genetic analysis of the trees. This is done free of charge, provided the university receives two specimens of each variety. Nobody is allowed to privatise the varieties; the agreement with the university stipulates that no one can claim plant breeders' rights on improved versions based on material from the varieties documented by the project. Otherwise, recipients of propagating material from the project can do as they wish, and it is an explicit goal of this project to offer to the general population the varieties freely and for free. With one of the apple varieties, four cider producers have expressed interest in making cider based on it, but these producers will have to sign a material transfer agreement where they agree not to exclude others from using the material if they want to receive the variety after the analysis has been completed.

As Mr Loma sees it, one of the main achievements of the orchard project is to show how people can work with and maintain these varieties, and to communicate their usefulness to the surrounding population. Information dissemination is central to the project, including assisting people as to how their own varieties can be maintained. The talks and discussions contribute to greater awareness among both the general public and the local administration. However, these events are attended mainly by elderly people, who usually understand these issues and might already know or be interested in the local varieties. To attract the younger generations as well, consideration is being given to arranging fruit-tasting sessions where people can compare local varieties with the fruit normally found in supermarkets.

The challenge for the project in the next phase is to map and multiply the remaining local fruit varieties as well as to make sure these varieties are being used. Ensuring that the associated traditional knowledge is being kept alive is an important part of this. The distribution of saplings and propagation material might run into legal difficulties; on the other hand, distributing the varieties by providing saplings has always been included as an objective in project proposals, so the provincial government should be well aware of this and Mr Loma does not expect objections. If the varieties in question were registered, this might solve the legal problems, but the Basque Seed Network firmly rejects any form of registration that would imply privatisation of the varieties in question.

Mr Loma also sees the beginning of a larger project, and talks about expanding have been started with the provincial government. The aim is a bigger nursery at the province level as well as smaller projects at the local level that can conduct surveys, analyse the results and then multiply the varieties and return them to the local communities.

Views on the network

The views of a government representative

While acknowledging that the seed network has had some success, Mr Samaniego Gil sees this success as modest and regards the impact the network has had to date as small. Because so few farmers have felt any impact, and because, as he sees it, the network is not so relevant for what he calls 'professional' farmers, he finds it surprising that the farmers' union defends it. In his opinion, the official position of the union is a political position and not the result of member considerations, especially as regards large-scale farmers. He sees the main impact of the network as being related to local produce and the 'slow food' movement, but he also recognises that there has been a certain impact on the local government. To some extent the network has succeeded in changing the attitudes of government representatives. Network members are in contact with the environmental department of the local provincial government of Alava, some local parish councillors and mayors and these people are listening to what the network has to say.

Even though he feels the network has had only limited success, Mr Samaniego Gil thinks the impact will increase in the medium-term as consumers become more aware of the issues the network works on. He also thinks that the situation would have been different for the involved farmers if the network had not existed. These farmers probably feel more secure today, knowing that the seed network is in operation and defending what the farmers see as their right to save and exchange seed. Because the network is a relatively young organisation, he also thinks that more time will be needed to achieve greater impact.

Mr Samaniego Gil is worried about the seed-health aspect of seed exchange, especially with vegetative reproduction. But although he thinks this might become a serious problem if there is no control, he also feels that it can be resolved, for example, by instigating some sort of control of seed production. Decisions here should be taken by society as a whole, he feels, since the possible negative impacts of what he refers to as 'bad seed' could have consequences for consumers as well.

The views of a farmer

Ainhoa Iturbe Suberbiola from Aristieta Baserria in Bizkaia Province has been involved in the seed network since the beginning. She attended the start-up meeting in 1996, and when the network conducted its first seed survey she was one of the farmers it collected seed from. Since then she has also received seed from the network. Although she has been involved in their activities since the start, Ms Suberbiola does not attend the network meetings herself. The farmers' union has a representative in the network and she prefers to get information from this person.

The Basque Seed Network consists of different groups and individuals, not solely of farmers. In Ms Suberbiola's view, this means that it is not always adjusted to the situation on the ground: she feels that farmers should be more involved. As she sees it, scientists and people from other backgrounds might not know enough about the practical running of a farm and farm work, so active participation of farmers in the network and more direct contact with farmers is therefore necessary. Lack of communication can be seen as one of the network's problems. Among the core group there are no full-time farmers. Getting more farmers actively involved is a complicated issue, however, since many farmers do not feel they have time to spare for such involvement.

Ms Suberbiola thinks that, in the short-term, communication with farmers can be improved and farmer involvement increased through the farmers' union. It is difficult to get farmers to attend meetings, and this problem has been discussed in the network for years. Even if farmers do not attend the meetings, however, coordination and communication must be improved and specific initiatives should, in Ms Suberbiola's opinion, be coordinated with farmers. Still, she feels that this is a two-way-problem and farmers need to become more engaged in the issue of maintaining the local diversity. But for many farmers the maintenance of varieties and seed saving is something they do not have the time to do or worry about because there are so many other problems. It has been discussed whether the seed network itself should be a farmers' group but that would go against the history and set-up of the initiative as the founding members were not farmers.

Despite the lack of farmer involvement, Ms Suberbiola sees the work done by the network as important and necessary. She also appreciates how the network can provide her with seed from new varieties. In her opinion, *in situ* conservation is very important, and she sees awareness-raising as a vital tool for getting more farmers to cultivate local varieties. Information dissemination should focus on the advantages of local varieties; and in connection with this she notes the importance of ensuring access to seed. In her view, seeds from local varieties should be as readily accessible as other types of seed.

One of the obstacles that needs to be overcome to increase the use of local varieties is, as Ms Suberbiola sees it, the widespread belief that it is easier and better to buy seed because maintaining varieties and saving seed requires a lot of time and effort. To promote the use of traditional varieties in her area, a local network of about 20 organic farmers that she belongs to is trying to organise a system of seed saving whereby they delegate responsibility for maintaining certain varieties to certain farmers and then share the seed. This is very similar to what the Basque Seed Network does on a larger scale, so it would be an excellent idea for this smaller network to coordinate with the Basque Seed Network. There has been some communication, but the goal is for greater contact.

In Ms Suberbiola's opinion, the main achievement of the Basque Seed Network is that varieties that otherwise would have been lost have been saved. In addition, the network has succeeded in raising awareness of this issue in the local administration and government as well as among farmers. She also sees the work done by the network with regard to the legal situation as important. As she herself has always cultivated as much diversity as possible, the network has not affected her practice in that aspect but it has been important for her to get access to seed and assistance from the network. Because she feels responsible for maintaining local varieties, she appreciates the activities of the network: it means that such work does not rest solely on her shoulders, and she feels she can worry less about it. In her opinion, the main reason for the network's success is the motivation and enthusiasm of the people involved as well as the amount of work that they have put in.

Ms Suberbiola is aware that it is illegal to share seed from unregistered varieties, but she has been cultivating local varieties by saving and exchanging seed for a long time and intends to keep on doing this, despite the restrictive legislation. Because there are so few people who save and exchange seed, these activities have not run into problems in the Basque Country so far. However, Ms Suberbiola thinks this would change if more people were to begin engaging in these practices. She knows of one person who wanted to set up a nursery for local varieties but did not dare because of the possible legal problems.

So far, the Basque Seed Network has been distributing seed only in small quantities and on a small-scale. Those involved feel that their activity will be tolerated as long as it is on such a small-scale – which is in line with what is being said by the seed department about how it can turn a blind eye to the network as long as the network does not expand its activities. On the other hand, some farmers engaged in the maintenance of traditional varieties find it difficult to get access to the amount of seed they need. This is a problem especially for farmers who produce seed for sale, since it takes a long time for them to multiply the amounts needed for production. Ms Suberbiola feels that the seed legislation should allow for the exchange of seed from local varieties without the need for registration and that the Basque government, especially the agricultural department, should provide the network with more funding and give them due recognition for their contributions to the conservation and sustainable use of plant genetic resources.

Concluding remarks

Seed saving and exchange still go on in the Basque Country, but, as elsewhere, these are practices that are rapidly declining due to many factors – an ageing farmer population, adverse legal restrictions and the promotion of industrial hybrid varieties by farm extension services and supermarkets alike. On the whole, however, the Basque Seed Network can stand as an illustration of how relatively small initiatives can have an impact on the conservation of plant genetic resources for food and agriculture and the implementation of Farmers' Rights as defined in the Plant Treaty. Through its documentation of local varieties and efforts to make sure these varieties are being cultivated, the network has been making substantial contributions to the *in situ* conservation of the remaining plant genetic diversity in the Basque Country. Distributing seed to interested individuals is an important part of this, but the lack of legal space for the sharing of seed poses a barrier to their work. The work of the seed network is restricted by what its members perceive to be the need to operate with small quantities only and that also limits to what degree the network can help farmers with seed.

The network can circumvent the law as long as its operations remain small-scale. In many ways the legal situation therefore creates a sort of 'glass ceiling' for the network's initiatives. As some members see it, what the network might need is somebody who could challenge the system by distributing seed from local varieties in greater quantities, both to improve the access to this type of seed and to draw attention to the cause. However, that could prove risky, with legal ramifications for those involved.

For many farmers, the lack of time and resources is a continuing challenge and can explain their minimal attendance at network meetings and why many farmers have stopped cultivating traditional varieties. The time and effort required are often decisive factors when it comes to deciding which crops and varieties to grow. Maintaining local varieties can be time-consuming – especially if access to the amounts of seed needed is problematic, and it can take many seasons to produce enough seed to grow crops for sale. If provided with the legal space to distribute seeds in larger amounts more freely, the network could make it possible for more farmers to cultivate local varieties. However, given the current legal situation, the network focuses on maximising their opportunities and trying for as much impact as possible.

The legislative situation and the restrictions placed on the sale and exchange of unregistered varieties is not unique to Spain. Thus, the experiences of the Basque Seed Network and how it has managed to operate in a challenging context is something other initiatives can learn from. In most industrialised countries, and to an increasing extent in developing countries as well, seed laws restrict farmers' possibilities to maintain traditional varieties. Under such circumstances, circumventing the law may be necessary in order to maintain what is left of agricultural biodiversity. As in the Basque Country, distribution on a deliberately small scale may prove the only way to avoid legal difficulties.

In situ conservation is seen as central by the members of the Basque Seed Network. That is why the network has focused on documenting and distributing local varieties as well as raising awareness around the issue. If diversity in local and traditional varieties is to be maintained these varieties must be grown, not least because actual use and cultivation are crucial to the continued adaptation of the varieties to local environments and climates. With the climate undergoing change this has become a particularly vital concern, relevant in all parts of the world. Access to seeds is integral to this and as Ms Suberbiola puts it, *seeds are the basis of life*.

Notes

1   All individuals mentioned in the text were interviewed in December 2008 in the Basque
     Country, Spain. Since then there have been some changes with regard to the network's
     most active members with new individuals taking over for some of the previously most
     active, but the network continues to thrive.
2   The authors spoke with several such farmers when visiting the market at that time.

5 Combining Farmers' Rights and plant variety protection in Indian law

Tone Winge, Regine Andersen and Anitha Ramanna-Pathak

This chapter tells the story of how India, through its 2001 Protection of Plant Varieties and Farmers' Rights Act (the PPVFR Act), has established the legal space necessary for farmers to continue maintaining their traditional varieties and practices, in addition to introducing plant breeders' rights. The PPVFR Act stands as the most far-reaching legislation to date with regard to establishing rights for farmers to save, use, exchange and sell farm-saved seed. Uniquely, it confers concurrent rights to breeders and farmers while at the same time recognising farmers as cultivators, conservers and breeders.

The PPVFR Act deals specifically with the rights of farmers and states that farmers have the right to save, use and exchange seed, as long as the seed in question is not sold in branded packages. In addition, the Act provides for the protection of various other aspects of Farmers' Rights, as this chapter will show. India is among the first countries in the world to have passed legislation granting Farmers' Rights, providing inspiration for stakeholders involved in similar legislation processes in several other countries.

Agriculture and seed production in India

Despite the fast pace of modernisation, agriculture is still important to India's economy. More than half of the work force is employed in agriculture, and the sector stands for about 17 per cent of GDP.[1] Crop production is vast, with a wide range of grains, fruits, vegetables, oil seeds, cotton, rubber, spices, sugar cane, tea, jute and tobacco. The great majority of India's farmers depend on farm-saved seed, and the national government has initiated several programmes to increase the seed replacement rate in order to boost production.[2] The Seed Act of 1966 is in the process of being replaced, with the stated goals being to 'create an enabling climate for growth of the seed industry, enhance the seed replacement rates, boost export of seeds and encourage import of useful germplasm and create a conducive atmosphere for the application of frontier sciences in variety development' (Ministry of Finance, 2012). Nevertheless, traditional, informal seed supply systems remain central in small-scale agriculture and are vital to the livelihoods of the great majority of Indian farming families.

India's Protection of Plant Varieties and Farmers' Rights Act of 2001

The primary objective of the 2001 PPVFR Act is to establish an effective system for the protection of plant varieties, the rights of farmers and plant breeders and to encourage the development of new plant varieties (Preamble).[3] Further, it recognises and protects the rights of farmers in respect of their contributions made at any time in preserving, improving and making available plant genetic resources for the development of new plant varieties. The act is intended to accelerate agricultural development in the country, protect plant breeders' rights and stimulate investment for research and development in the public and private sector for the development of new plant varieties. Finally, the act seeks to facilitate the growth of the seed industry in the country in order to ensure the availability of high-quality seeds and planting material.

One element that distinguishes India's PPVFR Act from most other legislation on plant breeders' rights is precisely the specific inclusion of Farmers' Rights. Not only are Farmers' Rights featured in the name of the Act, there is also a separate chapter devoted to them. The act contains nine specific rights (Bala Ravi, 2004; Ramanna, 2006):

- *Rights to Seed*: The PPVFR Act gives farmers the right to save, use, exchange or sell seed in the same manner as they were entitled to before the Act (Article 39). However, the right to sell seed is restricted, as farmers may not sell seed of protected varieties in branded packages. The legal space for farmers in this regard is nevertheless much broader than in other legislation on plant variety protection and can be seen as a very good way to realise the rights of farmers to save, use and exchange seed while also protecting the rights of breeders.
- *Right to Register Varieties*: Farmers as well as commercial breeders can apply for intellectual property rights over the varieties they breed (Article 39). The criteria for registration of farmers' varieties are also similar to those for breeders (genetic distinctness, uniformity, stability), but importantly, novelty is not a requirement. This possibility of obtaining intellectual property rights over farmers' varieties is a unique aspect of India's law. A 'farmers' variety' is defined as 'a variety which has been traditionally cultivated and evolved by farmers in their fields; or is a wild relative or landrace of a variety about which the farmers possess common knowledge' (Article 2.L).
- *Right to Reward and Recognition*: The Act provides for the establishment of a National Gene Fund through which farmers who have played a role in the conservation of varietal development of plants can be recognised and rewarded (Article 45). The fee collected from breeders who are required to pay for benefit sharing is to be deposited in the National Gene Fund, and the money collected under the Fund can be used for support and reward to farmers engaged in conservation.
- *Right to Benefit Sharing*: The centralised National Gene Fund is intended to facilitate benefit sharing (Article 26). The Protection of Plant Varieties and Farmers' Rights Authority that oversees implementation of the Act is required

to publish the registered varieties and invite claims for benefit sharing. Any person or group of persons or firm or governmental or nongovernmental organisation can submit claims to benefit sharing.

*   *Right to Information and Compensation for Crop Failure*: The Act provides that the breeder must give information about expected performance of the registered variety (Article 39.2). If the material fails to perform, farmers may claim compensation under the Act. This provision is intended to ensure that seed companies do not make exaggerated claims about the performance (yield, pest resistance). It enables farmers to apply to the Authority for compensation if they suffer losses due to the failure of the variety to meet the targets claimed by seed companies.
*   *Right to Compensation for Undisclosed Use of Traditional Varieties*: If it can be established that the breeder has failed to disclose that the source of a variety belongs to a particular community, compensation can be granted through the Gene Fund (Article 28). Any NGO, individual or government institution may file a claim for compensation on behalf of the local community if the breeder has not acknowledged use of the traditional knowledge or resources of the community.
*   *Right to Adequate Availability of Registered Material*: The breeder is required to provide adequate supply of seeds or material of the variety to the public at reasonable prices. If the breeder fails to do so after three years of registration of a variety, any person can apply to the Authority for a compulsory licence (Article 47). Such compulsory licences revoke the exclusive right given to the breeder and enable third parties to produce, distribute or sell the registered variety.
*   *Right to Free Services*: The Act exempts farmers from paying fees for registration of a variety, for conducting tests on varieties, for renewal of registration, for opposition and for fees on all legal proceedings under the Act (Article 44. See also Bala Ravi, 2004)
*   *Protection from legal infringement in case of lack of awareness*: Recognising the low literacy levels in India, the Act provides safeguards against innocent infringement on the part of farmers (Article 42). Farmers who unknowingly violate the rights of a breeder shall not be punished if they can prove that they were not aware of the existence of breeder's rights.

Article 3 of the PPVFR Act establishes the Protection of Plant Varieties and Farmers' Rights Authority (the Authority). The Authority shall make decisions by majority voting. It is the duty of the Authority to promote the development of new plant varieties and to protect the rights of farmers and breeders (Article 8). Measures it may employ in that connection include the registration of extant varieties; developing characterisation and documentation of registered varieties; documentation, indexing and cataloguing of farmers' varieties; compulsory cataloguing facilities for all plant varieties; ensuring that seeds of registered varieties are available to farmers and providing compulsory licensing if necessary; collecting statistics for plant varieties and ensuring the maintenance of the National Register of Plant Varieties.

The Authority is to consist of a Chairperson and 15 members, one of whom is to be a representative from a national or state level farmers' organisation. In addition, one member is to be a representative from a national or state level women's organisation working on agricultural issues. The seed industry and various government institutions are also to be represented. On the other hand, the farmers' representative as well as the seed industry representative and the women's organisation representative are to be nominated by the central government. It is indeed praiseworthy that such organisations are represented, but measures need to be put into place to ensure that the organisations are given the space to democratically select the representatives. Article 3 also states that the Chairperson shall appoint a Standing Committee to advise the Authority on Farmers' Rights issues. This committee is to consist of five members, one of whom is to be a representative from a farmers' organisation. Both provisions help to promote the participation of farmers in decision-making processes related to the management of plant genetic resources for food and agriculture.

The process of enacting India's PPVFR Act[4]

The passage of the PPVFR Act came as the result of substantial debate and five revisions. Whereas it was the demands of the seed industry for plant breeders' rights that had initiated the process, the chapter on Farmers' Rights was added as a result of pressure from NGOs.

Prior to the PPVFR Act, agriculture had not been included in the system for intellectual property rights in India and there had been no plant variety protection. The common heritage principle had dominated: crop genetic resources were viewed as part of a common human heritage that could not be 'owned' by anyone, and farmers had been free to use, share and exchange seeds.

However, after the New Seed Policy of 1988 allowed private companies to enter the seed sector, the first demands for plant breeders' rights were voiced. The Seed Association of India, which was formed in 1985, was one of the first organisations to promote plant breeders' rights in India. Then in January 1995, India became a member of the World Trade Organization (WTO), which entailed the obligations of the TRIPS Agreement – including the requirement for plant variety protection. Under the TRIPS Agreement, WTO members are obliged to protect new varieties of plants, whether by patents or by an effective *sui generis* system or by any combination thereof (Article 27).

Initially, the public sector in India had been opposed to plant variety protection, partly fearing that private companies might take unfair advantage of breeding material developed by the public sector under such a system. Later, as a result of the changing role of the private sector and the changing relationship between the sectors, the stance of the public sector shifted (Seshia, 2001). However, many NGOs and farmers' organisations protested against the introduction of plant breeders' rights, arguing that such legislation would recognise only the innovations of breeders and commercial firms and not those made by farmers and local communities.

With this debate as the backdrop, in 1993/1994 the Indian government created a draft bill on plant breeders' rights. This first draft caused considerable controversy despite the government's attempt to take the views of various stakeholders into account. The provisions of the draft were based on UPOV '91, but there were also formulations on community rights and Farmers' Rights – including the rights of farmers to save, use and exchange seed and propagating material as well as benefit sharing. However, this first draft did not mention the registration of farmers' varieties.

When the draft bill was opposed by both involved NGOs and the national seed industry, the government began a revision process. A second draft was presented in 1996 and a third in 1997. The words 'Farmers' Rights' had now been included in the title, but the NGOs in question still argued that the draft bill did not provide enough protection to farmers. The main points of concern for these organisations were what they perceived as too vague benefit-sharing provisions, no farmer representation in the Authority and the continued lack of a system for registration of farmers' varieties.

A fourth draft that tried to accommodate the interest of various stakeholders was presented to the Indian Parliament in 1999 and sent to a Joint Parliamentary Committee. This Committee then travelled around the country to gather the views of farmers' organisations, NGOs, scientists and the seed industry. Thereafter, the bill was redrafted – and in 2000 a new version that sought to incorporate the various views was introduced. As a result of this process, a separate chapter on Farmers' Rights was now included and a system for registration of farmers' varieties was introduced.

This fifth version was generally accepted by the major stakeholders. The national seed industry had come to see the system that was introduced for registration of farmers' varieties as something that would strengthen their position regarding intellectual property rights; paving the way for plant breeders' rights in India. The involved NGOs accepted this fifth version because of the mechanism it provided for protecting farmers' varieties along with breeders' varieties. Although some farmers' organisations (mainly representing large-scale farmers) had been consulted by the joint parliamentary committee, farmer participation cannot be termed comprehensive: it was mainly NGOs, public-sector institutions and private companies that participated in the process and had their views reflected in the final draft. On 30 October 2001, the PPVFR Act was passed.

Impact of the PPVFR Act

India's PPVFR Act has tried to uphold the legal space for farmers to save, use, exchange and sell farm-saved seed. In terms of enacting a law, it has been highly successful. Nevertheless, the ambitions were much higher: to protect Farmers' Rights as recognised in the nine rights that it provides for. Given these high ambitions, there has been discussion as to how successful the Act has been.

While opinion is divided on the implications of the law, with some seeing it as progressive and others questioning its real impact on farmers as compared to its ambitions, most observers would agree that the process of developing the legislation has provided important lessons. First, it was probably the first time the rights of farmers received such wide attention and stirred such debate both within and outside the Indian Parliament. Second, the government was forced to initiate wide-ranging dialogue with various stakeholders, as it could not manage to pass the legislation without their demands being met. This resulted in the Joint Parliamentary Committee travelling across the country and ultimately producing a draft that included a separate chapter on Farmers' Rights.

India's PPVFR Act can be seen as an attempt to develop a multiple rights system. While the Act seeks to distribute ownership rights in a fair and equitable manner, the assigning of multiple rights could pose several obstacles to useful utilisation and exchange of resources. The attempt to extend plant variety protection as is done in the Act may lead to a number of claims over a variety. If, for example, a plant breeder applies to register a new variety and breeding that new variety requires use of material (variety) over which any representative for another stakeholder group holds a right they can enforce, the plant breeder must bargain and pay for its use. Four stakeholder groups are listed in the Act: breeders working for private companies, breeders working for public institutions, NGOs, farmers or farming communities – and with each holding one of the types of rights, there are 16 possible combinations of stakeholders or claims (Ramanna and Smale, 2004).

The situation becomes even more complex when we consider that in many cases more than one type of material is used to breed a new variety and that additional actors may previously have gained ownership rights over them so that overlapping claims result (Ramanna and Smale, 2004). Some actors are also less capable of asserting their rights and will be left out – and these are likely to be small-scale farmers and their communities. For these stakeholders, the costs of claims may be too high.

However, as yet no systematic assessment has been conducted to evaluate how this multiple rights system works in practice and what challenges it may entail. Also, the number of farmers' varieties registered thus far is limited. For 2010/11, out of a total of 642 applications, 30 applications were received in the category of farmer's variety. Of these 24 were for rice varieties, four for kidney bean varieties and two for ground nut varieties (PPVFR Authority Annual Report, 2011). Perhaps this low interest in registering farmers' varieties is a reflection of farmers' priorities and lack of information. It could also be argued that the most important aspect of the PPVFR Act for farmers is to uphold their rights to saving, using, exchanging and selling farm-saved seed of all kinds and that formal ownership of the varieties they cultivate is not seen as so important an issue. Historically farmers have shared their seed, and this may still be the shared norm in many communities.

Lessons learned

The most important lesson to be learned for other countries is that it is possible to uphold the rights of farmers to save, use, exchange and sell farm-saved seed, also within the framework of national legislation on plant variety protection. Like so many other countries, India is a member of WTO and has ratified the TRIPS Agreement. The country is thus required to 'provide for the protection of plant varieties'. With its PPVFR Act, India complies with the provisions in the TRIPS Agreement on the protection of plant varieties. Other countries in the same position should therefore be able to pass similar laws without neglecting their TRIPS obligations.

Another important lesson that can be drawn from India's experience is the need for wide-ranging consultations with various stakeholders in connection with establishing laws on Farmers' Rights. India's experience also reveals the importance of carefully designing the stakeholder dialogues. The outcome should not be a situation where the law aims to satisfy the most vocal actors, without fully heeding the implications for its target group – the farmers themselves. Involving individual small-scale farmers who may not be represented by farmer lobbies and ensuring that various groups of farmers are represented by legitimate representatives, could prove valuable in the future. It is also important to include in the discussions those stakeholders that may oppose certain aspects of Farmers' Rights. During the consultations in India, seed industry representatives stated that they would only support exchange of seed, not any right of farmers to sell seed, because they felt this would benefit the middlemen and not the farmers. The views of the seed industry need to be taken into account, and these might help to ensure that the farmers themselves benefit from the inclusion of Farmers' Rights in legislation.

Another lesson relates to the harmonisation of laws and bodies in India. Improving coordination between the PPVFR Act and the Biodiversity Act is an enormous task in itself, let alone ensuring linkages between these acts and agricultural policies; for example, there is currently discussion on how the new Seed Act may influence the implementation of the PPVFR Act. Adding to the complexity is the fact that the Protection of Plant Varieties and Farmers' Rights Authority is tasked with implementing the different systems of breeders' rights and of Farmers' Rights. In addition, there are various institutions working to promote India's agriculture and development that benefit farmers. While each organisation usually focuses on one particular aspect, oversight of the total picture is not entrusted to any one body. The need for harmonisation of laws and institutional arrangements, which provide a clear division of labour, is thus an important lesson in this regard.

Furthermore, we can note that massive and enduring advocacy may be required in order to succeed with demands for Farmers' Rights in the context of the development of legislation on plant variety protection.

The Indian case provides important lessons for other countries as regards including provisions on Farmers' Rights in their national legislation. It also illustrates the complex and contentious issues that must be tackled if the legal space to save, exchange and sell farm-saved seed, as well as the other aspects of Farmers' Rights, are to be realised.

Notes

1   According to the CIA World Fact Book, 52 per cent of India's labour force works in agriculture, while the sector contributed 17.2 per cent of GDP (see www.cia.gov/library/publications/the-world-factbook/geos/in.html the estimates are from 2009).

2 According to the Government of India's official Economic Survey 2010–11, more than 80 per cent of Indian farmers rely mainly on farm-saved seed. As the central government is worried about the low seed replacement rate, it has initiated various programmes and schemes (www.indiabudget.nic.in/budget2011-2012/es2010-11/echap-08.pdf).

3 See also presentation of the objectives by the Protection of Plant Varieties and Farmers Rights Authority www.plantauthority.gov.in/pdf/FAQ_New.pdf

4 This section is based on Ramanna, 2006.

Part III

Success stories from the protection of traditional knowledge

6 Cataloguing potato varieties and traditional knowledge from the Andean highlands of Huancavelica, Peru

Maria Scurrah, Stef de Haan and Tone Winge

In 2006, a potato catalogue from Peru broke new ground in the documentation of agricultural biodiversity and traditional knowledge. The *Catálogo de Variedades de Papa Nativa de Huancavelica – Perú* (Catalogue of Native Potato Varieties from Huancavelica, Peru) was published by the International Potato Centre (CIP, Centro Internacional de la Papa) and the Federation of Farmer Communities of the Department of Huancavelica (FEDECCH, La Federación Departamental de Comunidades Campesinas de Huancavelica) under the coordination of Stef de Haan. It documents 147 native potato varieties and the related traditional knowledge of Quechua farmers in the Huancavelica region of Peru. When the catalogue was published in 2006, three years had passed since activities had been initiated; during that time, altogether eight different communities had been involved in this detailed and ambitious project.

Since documenting traditional knowledge is central to protecting such knowledge from disappearing and dying out, this catalogue constituted an important contribution to the realisation of Farmers' Rights in Peru. This chapter tells the story of how the catalogue was created and relates the views and experiences of the involved farmers as well as those of its other users, looking back on the process, the catalogue itself and some of the results.

Seeking farmers' views in the mountains of Huancavelica

Three years after the publication of the catalogue, representatives of the civil society organisation Grupo Yanapai sought to find out how the catalogue and its effects were viewed by the farmers who had contributed to it. One of the contributors they interviewed was Antonio Paitan Ccantu, a farmer from Pucara, Huancavelica, who cultivates a considerable number of native potato varieties.

Trudging uphill in the rain at the edge of potato fields and highland pastures in Pucara in 2009, representatives from Grupo Yanapai saw the family homestead of Mr Ccantu lying some distance below. A woman who was shepherding a flock and had taken cover from the rain under a blue plastic sheet said that Mr Ccantu was not in his house and pointed to a tiny figure at the bottom of the hill working inside a fenced field of pasture, despite the rain. Since another farmer engaged in the maintenance of potato diversity had just told them that they would certainly find Mr

Ccantu at home because a rock had injured his leg, they were somewhat surprised to hear he was out working. But the man working so vigorously indeed proved to be Mr Ccantu. Somewhat annoyed at being interrupted in his weeding, he instructed a small boy to keep an eye on the cattle and told him that he would be back shortly.

On the way up to his house he explained that a long steel pin had been inserted in his leg and that he would need a second operation soon. Difficult as it might be to believe, he was still working hard and walking well up the hill. Arriving at the farm he unlocked one of the buildings and revealed a spotless, modern kitchen that, unlike almost all the others in the area, was not black from soot but had a gleaming chimney and a window.

Mr Ccantu said that he very much liked the idea of cataloguing his potato varieties. The catalogue had made his work more visible and had led, among other things, to visits by various scientists interested in his varieties and the catalogue. He also appreciated how the catalogue had given him the opportunity to see which potato varieties were being grown in Huancavelica.

Creating a catalogue of native potato varieties

A catalogue like the potato catalogue from Huancavelica documents a collection of native varieties and the associated traditional knowledge, but it can also strengthen the living traditional knowledge. The best way to protect such knowledge, which is usually oral and often concerns practices that can be difficult to record, is to preserve it alive, through use. Through the process of creating a catalogue of potato varieties and distributing the end product to participating farmers and communities, as well as other stakeholders, a project like the Huancavelica potato catalogue can increase the awareness, use and exchange of varietal diversity and associated traditional knowledge. This in turn will contribute to keeping traditional knowledge alive and in practice.

Initially, the intention behind presenting the native potato varieties of Huancavelica in a catalogue was not clearly defined by the organisers, but the general idea was for the catalogue to serve as a type of baseline and field guide that could be helpful in the identification of varieties. When the project was first planned in 2003 there were no available examples of *in situ* catalogues. The main source of inspiration was therefore a catalogue from 2000 presenting a collection of native potato varieties from the gene bank of the PROINPA foundation in Bolivia (*Papas Bolivianas: catalogo de cien variedades nativas,* by Ugarte and Iriarte, 2000).

Various elements of an *in situ* catalogue and what it should (ideally) look like were discussed by field staff, farmers and CIP scientists. The most important aspects emerged as being explicit recognition of the communities and farmers who maintain this diversity; the inclusion of local ethno-botanical knowledge (nomenclature, histories and uses); DNA fingerprint for protection against misappropriation and future (re)identification; as well as text in the Quechua language, which is spoken by some 80 per cent of the people of Huancavelica and is the most common indigenous language in South America.

The first part of the documentation process involved identifying communities and farmer families who might take part in the project and contribute with their varieties and their knowledge. Taking into account their position along a north/south transect through Huancavelica, ethnicity, the distance to markets and the presence of native potatoes, a representative group of eight Andean communities was chosen. Within these communities, locally recognised farmer families and farmer groups who maintained high numbers of varieties were identified and asked to participate in the initiative. After the participants had been identified and had agreed to take part, on-farm trials for morphological characterisation and plant photography were initiated.

Figure 6.1 Leoncio Quinto with his copy of the Potato Catalogue
Source: Stef de Haan

Between 2003 and 2005, during two consecutive agricultural seasons, more than 20 farmer families and groups contributed their native varieties. These varieties were planted on the farmers' own lands, in plots containing five to ten specimens of each variety. Varieties normally grown in mixtures were now grown separately to enable individual evaluation. Each variety was morphologically characterised using the CIP descriptor list (see Gómez, 2000; Huamán and Gómez, 1994). Photographs were then taken of the plant, the leaf, the flower and the tubers.

For the molecular characterisation, DNA samples were extracted from the leaf tissue of all the potato varieties. These samples were used for the genetic finger-printing of accessions with microsatellite markers, which are also known as simple sequence repeats or SSRs (for more information about SSRs, see Ghislain *et al.*, 2004). Ploidy – the number of complete sets of chromosomes in a biological cell – was established through microscopy and flow cytometry, a technique for counting, examining and sorting microscopic particles suspended in a stream of fluid. What species the various potatoes belonged to was determined on the basis of their ploidy and morphological characteristics.

A series of ethno-botanical inquiries was undertaken to document traditional knowledge. Through these inquiries the names used for the different varieties were registered, together with information about the uses of the varieties (e.g. in specific dishes, for medicinal purposes or for freeze-drying) and any additional information regarding traits (e.g. with regard to storage or resistance) or traditions and myths the communities and farmer families might possess. Most of the key information came from households that cultivated particular varieties and older custodian farmers who enjoyed community recognition. These inquiries showed that such knowledge was often shared by both men and women but that women tended to be particularly knowledgeable about seed selection, seed storage and traditional cuisine.

In accordance with the Peruvian legislation on the protection of collective knowledge of indigenous communities related to biological resources (Law 27811), agreements of prior informed consent were signed by each of the communities and farmer families. Moreover, in a clause on the title page of the catalogue it is explicitly stated that the collective knowledge presented is the property of the participating farmer families and communities. This was done to avoid misappropriation. In order to create regional ownership, an agreement was also signed with FEDECCH concerning the protection of the collective knowledge presented in the catalogue and its co-publication with CIP. This institutional agreement was signed by the President of FEDECCH and the Director of CIP. The agreement stipulates that any revenues from sales of the catalogue are to be placed in a specific fund to support future publication of *in situ* catalogues.

In addition to the variety descriptions and the information about traditional knowledge, the catalogue contains other features. The participating communities and farmer families are presented in both text and photographs in the catalogue, as a way of showing recognition to them and their contributions to the conservation of potato genetic resources. For the sake of openness and transparency, the objectives of the catalogue are clearly stated and the process of how information was obtained is outlined. Also important is the visual aspect of the catalogue. Abstract concepts such as 'molecular fingerprints' are presented in a clear and attractive way, and colourful pictures accompany each presentation. After publication, the catalogue was distributed to all the participating families and an additional 30 copies were left for each community.

Satisfaction with the catalogue

All the farmers who were interviewed about the catalogue in 2009 answered, as Mr Ccantu had done, that they were very pleased about their varieties being documented in the catalogue. They liked the idea of making their knowledge available to a wider audience and many said that doing so made them feel proud. Many also pointed out that sharing this knowledge would protect it from being lost. There was agreement among farmers and scientists alike that the catalogue provided useful knowledge about which varieties were currently grown in Huancavelica, and this in itself was recognised as very valuable. Farmers can now compare the varieties listed in the catalogue with those they themselves are growing, to find out which ones they lack. For scientists, the catalogue represents a baseline to enable monitoring of varietal change or loss in the future.

Although Peru is recognised as the centre of origin of the potato and the country where the greatest diversity of wild and cultivated potatoes can be found, little is known about precisely which varieties are cultivated or where. For the scientists interviewed, an important aspect of the potato catalogue was to make this diversity visible while also giving the farmers credit for their maintenance work. Until recently, all available data on potato diversity in Peru came from the *ex situ* collections established in the 1970s and 1980s by CIP. For a long period after that, during the years of political violence (1980–2000), there were fewer collecting expeditions because of the dangers associated with travelling around in the countryside. The Convention on Biological Diversity, with its provisions on access and benefit sharing, also affected the mapping and collection of potato diversity in Peru. With the ratification of the Plant Treaty by the government of Peru in 2003, the possibilities of collecting have again opened up. However, no new potato collection expeditions have been undertaken as of 2012.

Farmers and scientists also agreed on the importance of the illustrations used in the catalogue and felt that the photographs were the key to the success of the catalogue. The farmers stressed how useful the photographs of the tuber shapes and colours, the flesh colours and the flowers and leaves were, and how wonderful they found them. For many farmers, looking at the photographs constituted 'reading' the catalogue – very few of them had spent much time on the actual text.

One farmer, Paulina Ortiz Palomino of the Dos de Mayo community, knew all the varieties and could say which ones she thought had the wrong names or where local names had not been listed, even though she is illiterate – thanks to the photographs and because her daughter-in-law had read to her. Farmers and scientists alike were so insistent on the usefulness of the good photographs that one might well wonder whether the 25 descriptors so laboriously developed by CIP were in fact necessary. The old adage 'a picture is worth a thousand words' certainly rings true with regard to this catalogue.

Among the stakeholders interviewed there was also general agreement on the likely usefulness of the catalogue in the future. Most children in these farming communities can now read, and the farmers thought that the catalogue might prove even more useful to their offspring than to themselves. But they also underlined that the catalogue would help adults to remember the different varieties. Some of the young farmers thought that many of the varieties would disappear from their fields and saw the catalogue as a way of conserving them while more optimistic farmers thought that the catalogue would help in maintaining such varieties because it would help them remember. Among the scientists the catalogue was seen as a record of what was in the fields at the time it was made, and they thought it would have many possible uses in the future. Indeed, one of the original intentions had been to record the existing diversity in the fields and the frequency of particular varieties at the time (total versus relative abundance).

However, both farmers and scientists felt that the catalogue was incomplete. Many farmers mentioned that they themselves grew varieties that were not in the catalogue. It is estimated that between 100 and 200 of the varieties grown in Huancavelica were not included – which means that only about half of the potato varieties actually cultivated in the area made it into the catalogue. Part of the explanation for this is incomplete data: for instance, if there was no photograph of a flower or tubers, the variety in question was not included. Nevertheless, the maintenance of potato diversity is a dynamic process, and the catalogue might need to evolve along with cultivation practices, for example with a new folder-based version to allow new pages to be added.

Exchange and movement of potato diversity

In recent years, many of the seed fairs held in Peru have given awards to the farmers who display the greatest number of varieties. As a result, some farmers have begun to collect varieties, hoping to win local, regional and national awards. In 2008, the 'year of the potato', 30 Peruvian farmers were chosen to receive a national award for their work in maintaining potato varieties, at an event sponsored by the Ministry of Agriculture as part of the official celebrations. Mr Ccantu and two other farmers who also took part in the catalogue project, Juan Ramos and Leoncio Quinto, were among these as all three of them grow a large number of varieties.

During his interview, Mr Ccantu was pleased to announce that in connection with the national award he had been able to exchange varieties with farmers from Huaráz, Huánuco and Cusco and that as a result he had brought home at least 50 new varieties. Moreover, Juan Ramos and Leoncio Quinto had exchanged varieties in Lima and had brought new ones home. This illustrates how varieties can move from region to region.

Farmers like Mr Ccantu, Mr Ramos and Mr Quinto are constantly bringing in new varieties to the area where they live because they travel to other regions in search of work. The influx of new varieties in this manner to Huancavelica is rapidly increasing. As a result of new roads, farmers now go to major towns and cities in search of work, even to places further away like Puno and Juliaca near the Bolivian border, and they often bring back varieties from these areas. According to the CIP gene-bank curator, René Gómez Zarate, many of the varieties found in Huancavelica can also be found in Ayacucho, Junín and Apurimac, and even Cusco.

The unrest and political violence that haunted Peru for over a decade was another factor that contributed to shifts in the variety composition of various areas. When farmers fled the political violence in the 1980s they abandoned their fields, often losing their varieties. These varieties were then later probably replaced by varieties from elsewhere. In some areas, as in the case of the Chopcca community, they managed to keep their own varieties because they were able to put up a fierce defence against both the Maoist *Sendero Luminoso* guerrilla and the national army.

Huancavelica has also suffered some variety loss due to frost. In February 2007, for instance, there were widespread frosts and many families lost potato varieties. After a frost the variety composition in an area will usually change. Frost-tolerant varieties become more prevalent but so may improved varieties which, although usually susceptible to frost, become widespread because they are sent as relief seed and are easily available in local markets. However, the earlier susceptible varieties are not necessarily lost. Some areas are always spared from the frost, and from here the varieties might slowly reach more fields through farmer-to-farmer exchange. In this way a new equilibrium is often established though dynamic seed-exchange networks.

The entire Andean potato farming system relies on exchanges between and within communities and regions. The wide range of varieties found in Huancavelica today is the result of farmers from ancient to modern times bringing in varieties from other areas and then selecting for adaptation, taste and use in their fields and homes. In colonial times, the mining centre of Santa Barbara in Huancavelica was a principal supplier of mercury for the Potosi mines in Bolivia, and the regular trade routes without doubt also influenced the flows of seed.

Such movement and exchange show how the collection of varieties grown in any one area is necessarily a mixture of old and new varieties, and how with time, new varieties may even come to make up the majority. When researching areas with a considerable amount of diversity it is therefore difficult to differentiate between places that are rich in diversity due to the diligence of farmers in bringing in varieties from other areas and areas that have maintained a status quo regarding varietal portfolios for a long time. In order to make comparisons and truly comprehend the temporal dynamics of varietal turnover and change, what is required is a baseline – as in the form of a catalogue.

Throughout its history CIP has used various techniques to identify duplicates and has narrowed down its original 13,000 accessions to a collection of around 4,000 separate entries. This is another illustration of the extent to which varieties travel widely in Peru and how, as a result, many of them are cultivated and known throughout almost the entire country. The fact that CIP staff collected the same varieties from different areas demonstrates that it is difficult to know where any variety originates, even if the variety in question is very widely grown in one particular area.

After the publication of the potato catalogue, the region of Huancavelica became known as a 'hot spot' of potato diversity. In this way, the catalogue managed to direct attention to a little-known and undervalued area.

Figure 6.2 Mixed potato varieties from Huancavelica
   Source: Stef de Haan

Naming complications

The exchange of varieties between communities and regions has also had an impact on how the varieties are named, and sometimes the same variety has been known by different names in different areas. One of the most popular potato varieties in Peru, *Ccompis*, is recognised as a Cusco variety. However, this variety is also very popular in Huancavelica – where it is considered to be a local Huancavelica variety and is called *Prescos*. The Huancavelica name may stem from a mispronunciation of the Spanish word *precoz* (early) as harvesting for this variety takes place relatively early. There used to be a trade route from Huancavelica to Potosi that went through Cusco, so this variety could have arrived in Huancavelica already during the colonial period.

The subject of vernacular names can be a complex issue, and it is one of the topics the farmers tended to bring up when asked about their views on the catalogue. The name of a variety is important, because it serves as the basis for transferring the associated knowledge from person to person and area to area. However, because of the oral character of the Andean culture, potato varieties have often changed names as they travelled from one region to another. Research conducted in the eight participating communities has shown that the most commonly used varieties are usually known by one name only, whereas less common or scarce varieties will frequently have many different names.

According to Mr Ccantu, all the names used in the catalogue are correct – but here it should be borne in mind that to a certain extent the project relied on his knowledge with regard to naming. However, the names were a source of irritation to many other farmers, especially to the farmers from Chopcca, Allato and Pongos, who protested that their varieties showed up in the catalogue under other names.

Juan Ramos claimed that when the names are changed from Quechua to Spanish, the varieties lose their identity. A farmer from Pongos insisted that the 'legitimate, original names' should be used in catalogues, and one of the scientists suggested that a meeting should be held where one name for each variety could be agreed on for the next catalogue.

In the catalogue itself the issue of multiple names is dealt with in a box called 'synonyms'. However, almost none of the farmers had noticed this small box, and even if they had, the names given there usually did not coincide with what they knew as the 'right' name. In Chopcca, the farmers recognised 71 varieties from the photographs, but even after reading the synonym box they concluded that only 20 of these had been catalogued under the right names. Nonetheless, the farmers did sometimes recognise that a variety may be known under more than one name.

The use of molecular markers

When multiple names for the same variety cause confusion, tools like molecular markers can be of help in sorting out duplicates. In an effort to make the molecular markers used in the catalogue project easier for the readers to understand and to connect the technique to the history and culture of Peru, such markers have been presented in the catalogue as a type of *Khipu*.

The *Khipu* was a system of knotted cords used in the Inca Empire to record information (see Urton, 2003). In the catalogue, the molecular markers are used as a graphic representation of a genetic fingerprint, with each of the 18 strings representing a microsatellite marker and each knot representing an allele (for more on this, see Stapleton, 2006). The position on the string is an indication of its size: the higher up on the string, the more base pairs it has. Allele frequency is shown by the use of colours.

Only one of the involved farmers interviewed in 2009 mentioned the molecular markers or Khipu – and he said they should be omitted from the next catalogue because they were incomprehensible. This has probably less to do with the way the molecular markers are presented than with the fact that most Andean farmers lack the underlying knowledge systems to make sense of such a technology.

Among the scientists the molecular markers were also controversial. Some thought they were one of the best things about the catalogue while others felt that they were 'out-dated' because the new set of molecular markers now uses 24 markers to be in accordance with the chromosome number of the potato.

Regardless of the particular technique used, however, the markers would probably remain difficult to understand for most farmers and other non-scientists without lengthy explanations. And molecular fingerprinting tools do change frequently. It was also mentioned that although the primers used are listed in the catalogue, it is doubtful whether any other laboratories would have access to them. However, that concern might reflect lack of knowledge about the issue since all the primers used are in fact publicly available.

The molecular marker technique used in the catalogue may prove useful for identifying duplicates, and CIP has selected and developed a kit of highly poly-morphic microsatellite markers toward this end. Documenting varieties by the use of molecular markers might also prevent misappropriation of local varieties.

Scientists from various universities and from Peru's National Research Institute found the molecular markers very useful, even more so than the morphological descriptors, and felt they would be essential to the rapid identification of varieties in the future. However, using this technique is relatively costly, and in at least one recent catalogue inspired by the Huancavelica catalogue the molecular markers have been omitted, due to the high costs *(Las Papas Nativas de Canchis* by Gutiérrez and Valencia, 2010).

One way of making the scientific information more accessible could be, as a professor from the Universidad Nacional del Centro in Huancayo suggested, to have a section called 'genetic information' in a separate square or box, where information about the species, ploidy level and the molecular markers would be given. As another solution, it was also suggested that there should be two versions if the catalogue is revised: a printed version for farmers and an electronic version with scientific information for scientists and breeders.

Use of the catalogue

As the farmers see it, the catalogue is mainly useful for comparing, remembering and knowing the different potato varieties. The only use they can envisage outside their own communities is for marketing purposes. One farmer thought he would travel with his potatoes and catalogue down to the area where maize is being grown, to barter potatoes for maize. He would use the catalogue to point out the varieties he had brought and everybody would be able to see their cooking characteristics prior to the exchange, without him having to boil the potatoes to prove that his varieties are not bitter.

Very few farmers thought that non-farmers outside their community would be interested in the catalogue. This indicates that most Andean farmers do not believe that outsiders value traditional farming or its importance for the conservation of genetic resources. They would probably be very surprised to learn that most copies have been sold to people in cities abroad who see the catalogue and its entries as proof that Peru is a country rich in agricultural biodiversity.

On the other hand, few Peruvian scientists and researchers have bought the catalogue, and many of them complained about what to them seemed a high price, US$30. Despite this the interviewed academics all agreed that the catalogue could be very useful for teaching, not only at the university level but also in secondary and primary schools. They also thought that the catalogue should be sent to politicians to convince them of Peru's potato biodiversity and help them realise the importance of formulating better policies to maintain this biodiversity.

Challenges and suggestions for new features

For the presentation of each variety in the catalogue there is a square with text in Quechua that contains ethno-botanical information from the farmers who provided the variety in question. Many see this as one of the catalogue's unique and most useful features. The background for the decision to have some text in Quechua was that several farmers and researchers pointed out that a catalogue of potato varieties from Huancavelica should be in the language used by the farmers there. In fact, however, none of the Quechua speakers or the bilingual speakers could read or understand these sections. Quechua is practised almost exclusively as an oral language, and most Quechua speakers who learn to read will learn to do so in Spanish. This approach has therefore not been as useful as it was hoped it would be. In fact, 50 per cent of the people interviewed three years after the catalogue had been published were highly sceptical of the use of Quechua because of the difficulty it poses in terms of reading.

For a new edition or new catalogues, using both Quechua and Spanish or publishing two different versions might solve this problem. One urban user from Lima stressed that the part in Quechua should definitely be translated since it was the most interesting part of the text. Like names, the stories and traditional knowledge related to each variety vary from farmer to farmer. Several people outside the farming community felt that it was more important to include this information in the catalogue than the scientific knowledge because it is this fragile oral knowledge that is most in need of conservation. The cultural richness through which farmers relate to their potatoes and their land is central in this context, and this richness is seen by many as being what needs to be captured in catalogues and showcased to the general public.

Many farmers wanted future versions of the catalogue to have a section with information about where and how each variety should be cultivated. Some varieties prefer a certain type of soil and climate, and the farmers wanted this information to be available. High-altitude Andean soils are extremely acidic and many native varieties are adapted to just such conditions. For such information to be included, of course the farmers who provide the varieties would have to be able to give it. Some farmers also wanted there to be information about the varieties' abilities to resist biotic and abiotic stresses, including pests and diseases. Information of this type had been included only for some key factors: frost and late blight/potato blight (*Phytophthora infestans*).

Because relatively little research has been done on Andean pests and diseases and even less on the response of various Andean potato varieties to such biotic stresses, it might be difficult to access this type of information. However, the fact that the farmers want such information shows that they are very aware of the usefulness of these traits and how varieties differ in their response to pests, diseases and climatic stresses. There is information available on some diseases as well as some new information on nutritional value, and it is hoped that in the future this information can be made available to farming communities. Whether a new catalogue would then be the most appropriate medium would need to be evaluated.

One of the most important challenges related to the maintenance of Peru's rich agricultural biodiversity is the question of poverty alleviation and how to improve the livelihoods of the farmers who maintain this biodiversity. Most of the farmers who participated in the catalogue project are categorised as poor, and statistics show that the areas where agricultural biodiversity is being conserved are characterised by extreme poverty. Maintaining a diversity of varieties is of course not in itself a cause of poverty, but it does require extra commitment from the farmers. It is difficult to say what impact accelerated environmental and socio-economic change will have on the on-farm conservation of agricultural biodiversity and how greater attention to benefit sharing might affect the lives of these farmers.

As it becomes easier and easier to access modern, improved varieties, incentives are needed to make it advantageous for Andean farmers to choose to maintain their potato heritage. Otherwise, Peru may face a rapid decline in the number of farmers committed to cultivating these varieties. In Colombia and Ecuador, potato diversity at the farm level has already decreased substantially. Many Peruvian women still prefer their own varieties because they are convinced these are superior and have the best cooking qualities, but to keep the traditional Andean agriculture and cuisine alive it will be necessary to lift the farmers out of poverty and make the maintenance of genetic resources advantageous and profitable. Mr Ccantu has been able to install a new kitchen and luckily could afford to have surgery – but not all Andean farmers are in the same position. This is something future catalogue projects should take into account, perhaps giving greater consideration to ways in which the farmers can benefit from maintaining and sharing their varieties.

Other catalogues

The potato catalogue from Huancavelica was the first *in situ* potato catalogue. However, two catalogues based on information and material from gene banks had earlier been published in Bolivia *(Catálogo de las variedades locales de papa y oca en la zona de la Candelaria,* by Merino *et al.*, 2004, and *Papas Bolivianas: catalogo de cien variedades nativas,* by Ugarte and Iriarte, 2000). As mentioned, especially the latter served as a source of inspiration when the catalogue from Huancavelica was created.

In 2006, the year the Huancavelica catalogue was published, a catalogue describing the variability of native potatoes from six communities in the Calca and Urubamba regions of Cusco, Peru, was also published *(Variabilidad de papas nativas en seis comunidades de Calca y Urubamba – Cusco,* by Cosio Cuentas, 2006). This catalogue offered only limited recognition to farmers as the custodians of the varieties presented and also differed from the Huancavelica catalogue in that no genetic fingerprinting or high-quality photographs were included.

When the catalogue from Huancavelica was published it received considerable attention, further spurring interest in documenting agricultural diversity and traditional knowledge. Catalogues increasingly came to be seen as important baseline tools. In 2008, as many as four potato catalogues were issued in Peru, and to a certain extent all were inspired by the Huancavelica catalogue. One of these, *Papas Nativas del Perú: catálogo de variedades y usos gastronomicos* (MINAG, 2008), was financed by the

Figure 6.3 Potato farmer Eulogio Escobar sorting some of the varieties that he grows
   Source: Stef de Haan

Peruvian Ministry of Agriculture and sought to take advantage of the growing market for potato diversity by focusing on cooking qualities. Two district-level catalogues were funded by the Swiss Agency for Development and Cooperation and created in collaboration with local NGOs and published by BIOANDES (*Diversidad de papas en el distrito de Pitumarca* and *Variedades de papas nativas y conocimientos campesinos: microcuenca Shitamalca, San Marcos, Cajamarca*, 2008a, 2008b). The catalogue *Pampacorral: catálogo de sus papas nativas* (Ugas, 2008) presents the family collection of farmer Julio Hancco Mamani from Cusco; staff from the National Agrarian University La Molina participated in the documentation process.

   The key difference between these other catalogues and the Huancavelica catalogue concerns the visual information for each variety, which in the other catalogues is generally restricted to a photograph of the potato without quality images of the plant, the flower or the leaves. In addition, the first generation of catalogues contained only basic content about vernacular names and primary uses, as opposed to the Huancavelica catalogue with its large body of information gathered from many different farmers.

In 2008, an innovative catalogue was released in Bolivia: *Catálogo etnobotánico de papas nativas: tradición y cultura de los ayllus del norte Potosí y Oruro* (Terrazas and Cadima, 2008). It presents ethno-botanical knowledge and gives clear recognition to communities, farmers and indigenous authorities. This catalogue also contains innovative features such as graphics indicating how the varieties benefit the growers (for instance, through sale, exchange, home consumption and freeze-drying) and references to the type of agricultural systems in which the varieties are grown (single varietal stands or mixtures).

And between 2009 and 2011, three new and important catalogues were published in Bolivia, Colombia and Ecuador (Iriarte *et al.*, 2009; Monteros *et al.*, 2011; Moreno *et al.*, 2009). Since then no new catalogues have been published in the Andean region. One innovative aspect of the 2011 Ecuadorian catalogue that may prove useful for farmers, consumers and nutrition experts is its presentation of the micronutrient content of each variety. The Ecuadorian catalogue in particular, has drawn heavily on the lessons learned from the Huancavelica catalogue and as a result provides due recognition to each of the communities and households that participated in the documentation process.

The Peruvian catalogue *Las Papas Nativas de Canchis* (Gutiérrez and Valencia, 2010) focuses on the processing qualities of native potatoes, based on the idea of marketing (naturally) coloured potato chips as a way of fostering the maintenance of diversity.

All these catalogues show the appreciable impact the early generation of *in situ* catalogues has had on various stakeholders within the Peruvian agricultural sector, most notably NGOs and the Ministry of Agriculture, as well as stakeholders in other countries of the region.

Concluding remarks

There is no doubt that the 2006 Huancavelica catalogue has captured the interest of farmers and researchers as well as other interested individuals and that it has succeeded very well in presenting the rich potato diversity of Huancavelica in an attractive manner. The catalogue was designed to appeal to farmers, scientists and urban users. By engaging farmers in the process and contributing to the conservation, use and sharing of traditional knowledge, the catalogue represents an important contribution to the realisation of Farmers' Rights as defined in the Plant Treaty. As the first of its kind in the Andean region, the catalogue has also inspired many other similar projects, some of them presenting potatoes from other regions.

The majority of the farmers interviewed for this chapter, especially the women, are illiterate. For them, the photographs proved to be the most important feature in the identification of each variety. Both scientists and farmers praised the photographs, calling them beautiful as well as useful. It was also regarded as important that the pictures would help farmers remember their varieties, both those they still cultivated and those that they had lost. All the farmers considered the photographs of the flowers and of the tuber flesh to be the most useful feature of the catalogue. Including pictures in this way is therefore something that initiators of similar projects should consider to make sure the end products can be used by, and benefit, the farmers as well as other stakeholders. If the associated traditional knowledge is to be kept alive, catalogues must be of practical use to farmers.

As names proved to be a contentious issue, new editions of the catalogue and new catalogues should include alternative names for the varieties, presented in a very visible form. In addition, more farmers should participate, to facilitate wider consensus about the use of different names, including synonyms. This should foster a feeling of ownership among the involved parties as well as promote the use of the catalogue.

Molecular markers are useful to the farmers only if they have access to laboratories and research results through researchers. The techniques used in any catalogue should therefore be chosen carefully, to ensure both that they are widely used and that they are unlikely to become out-dated and replaced soon. This is a challenge, as marker and genomic technologies are advancing and changing continuously.

In many ways, the catalogue from Huancavelica marked the beginning of increased recognition of farmers and their varieties in Peru. Of the catalogues published in the region since, the Ecuadorian one in particular has internalised lessons from the Huancavelica catalogue. From in-depth interviews with farmers it is apparent that cataloguing is a good way to conserve traditional knowledge. However, these farmers cannot be expected to sustain the maintenance and ongoing evolution of Peru's agricultural heritage alone: wider support from society is necessary, as are clear opportunities to improve their welfare. It is crucial for the durable conservation of crop genetic resources and the associated traditional knowledge that the involved farmers, most of whom are poor, are supported in their efforts to create better lives for themselves and their families. Simply recognising them, as the Huancavelica catalogue does, is not enough although it is an important first step. Applied innovative, replicable and sustainable models that can encourage and support farmers to conserve genetic resources on their farms need to be identified and disseminated.

7 A community registry in the Philippines

Regine Andersen, Teresita H. Borromeo and Nestor C. Altoveros

On the island of Bohol in the Philippines, local farmers' associations engaged in organic agriculture and farmer-led participatory plant breeding have taken steps to ensure that the rice varieties that they maintain and breed will continue to remain in the public domain, freely accessible to farmers. This was necessary, they felt, after new legislation on plant variety protection was adopted in the Philippines in 2002. According to this Act, farming communities and NGOs would have to register already existing plant varieties in order to prevent them from being taken over by other plant breeders and made subject to plant variety protection. In this way, the Act makes farming communities and NGOs responsible for establishing *prior art*.

Farmer breeders in the Bohol communities of Campagao, Zamora, Riverside, Cansumbol, Poblacion Vieja and Malitbog realised that action was required.[1] Since 1996, they had been developing their breeding activities combined with organic agriculture, with great achievements in terms of new varieties that were better adapted to local environmental conditions and which were more nutritious and filling as well as tastier. Since some of their produce could be sold on the market and command better prices, the financial situation of many of the farmers had also improved.

Mr Cisenio Salces, farmer of the Campagao community in Bilar, Bohol, is among the best-known farmer breeders of Bohol. In 2002 he had already developed 24 breeding lines/varieties, several of which were highly popular among local farmers and elsewhere in Bohol. In fact, the authors of this chapter were served delicious red rice of a variety bred by Mr Salces at an organic restaurant in Tagbilaran, the capital of Bohol, during our field visit. This surprise underlined the importance of the breeding work carried out by Mr Salces and his colleagues. Many farmers find their way to Mr Salces' fields, visit him and ask for seeds of different varieties. They stay in touch and share their experiences – and thereby also their knowledge on seed selection and breeding.

With the new Act on plant variety protection, however, the farmers feared that they could not continue freely sharing their seeds and knowledge. The more they shared the greater the risk of misappropriation would be, as long as they did not take measures to establish *prior art*. This is why they decided to establish a community registry of the rice varieties they maintained and bred, together with the knowledge related to their properties, growing conditions and use.

Figure 7.1 Farmer breeder Mr Cisenio Salces in front of rice fields in Campagao, Bilar
Source: Regine Andersen

The farmer breeders in Bohol had a difficult starting point when they first took up their rice-breeding activities. Most of the traditional varieties were no longer available, and only very few formally released varieties were accessible. The core challenge was to enhance the genetic base of rice in the area through breeding work. Farmers' knowledge concerns not only older varieties and practices but also the innovations they carry out in their daily work, in selecting the best material for further propagation and varietal improvement. Such knowledge is crucial to the role of farmers in maintaining and enhancing crop genetic diversity. This is an important reason why this case from the Philippines can be presented as a success story of the protection of traditional knowledge. Moreover, this story shows how it is possible to protect the knowledge in such a way that it remains in the public domain and can be freely shared among interested farmers.

The farmers in Campagao, Zamora, Riverside, Cansumbol, Poblacion Vieja and Malitbog are all organised in farmers' associations that have worked closely together with the Philippines-based NGO Southeast Asia Regional Initiatives for Community Empowerment (SEARICE). The support of SEARICE has been instrumental in developing the farmers' approach to establishing *prior art* and ensuring that the varieties they maintain and breed remain freely accessible for all. In particular, SEARICE support was central in ensuring that the legal requirements for the community registry were met, to truly establish *prior art* with regard to the varieties important to the farmers.

SEARICE and its involvement in Farmers' Rights

SEARICE is a Philippines-based NGO operating throughout Southeast Asia and focusing on Bhutan, Cambodia, Laos, the Philippines, Thailand and Vietnam. In partnership with civil society organisations, government agencies, academic research institutions and local governments, it works to promote community-based conservation, development and sustainable use of crop genetic resources.

Established in 1977, SEARICE is aimed at facilitating the empowerment of farming communities as managers of crop genetic resources, thereby ensuring farmers' access to, and control over, these resources while developing economically viable and sustainable agricultural systems. It seeks better recognition and strengthening of farmers' initiatives, at the local, national, regional and international levels. The organisation conducts policy research in particular on issues affecting farmers' access to, and management and control of, crop genetic resources and local seed systems. It advocates policies that recognise, support and strengthen community initiatives in conservation, development and use of crop genetic resources. Furthermore, it seeks to develop the capacities of its partners for policy advocacy.

SEARICE was deeply involved in the negotiations that led to the Plant Treaty at the FAO during the 1990s, in particular with regard to the provisions on Farmers' Rights. Former director of SEARICE, René Salazar, was a driving force in the Philippine delegation. Since then, the organisation has continued as an active participant in sessions of the Governing Body of the Plant Treaty as well as at the Conferences of the Parties to the Convention on Biological Diversity.

In the Philippines, SEARICE has been actively engaged in policy developments regarding plant variety protection, access and benefit sharing related to genetic resources, and Farmers' Rights – in addition to its community-based activities for conservation and development of crop genetic resources in various areas of the country. In 1992, SEARICE established the Community-based Native Seeds Research Centre as a non-profit organisation. Its flagship programme was on the conservation and development of community plant genetic resources, aimed at the collection, conservation, research, development and utilisation of crop genetic resources, in partnership with farmers.

SEARICE was also one of the coordinating organisations of the Community Biodiversity Development and Conservation Programme (CBDC), an initiative for understanding and strengthening farmers' systems of conservation and development of crop genetic resources as well as of maintenance of biodiversity in general in Africa, Latin America and Asia. Whereas the CBDC Programme started in 1994, its Asian part was merged with the Biodiversity Use and Conservation Asia Programme (BUCAP) in 2006 into the CBDC-BUCAP, a still on-going programme coordinated by SEARICE and implemented in Bhutan, Laos, the Philippines, Thailand and Vietnam. The programme recognises the close linkages between biodiversity conservation and local sustainable development and seeks to promote local control of resources and strengthen the autonomy of farmers.

The CBDC and later CBDC-BUCAP projects in the Philippines are located on Bohol and Mindanao. They are aimed at strengthening farmers' capacities for managing their crop genetic resources and securing their local seed systems through conservation, crop improvement and sustainable utilisation. The collaboration between farmer groups and research institutions/policy-makers is a central target and is seen as instrumental for crafting programmes and policies that can support and strengthen farmers' and local communities' initiatives on conservation, development and sustainable use of crop genetic resources. An important element of CBDC-BUCAP has also been to broaden the genetic base in farmers' fields for rice, maize, yams, sweet potato and cassava, by assisting farmers in breeding crop varieties suited to local conditions. *In situ* management of crop genetic resources is a central component of the projects, which seek to enhance farmers' knowledge and skills in participatory plant breeding and provide them with better access to a diversity of crop genetic resources.

Conservation efforts and farmer breeding in Bohol prior to the Community Registry

SEARICE began its engagement in activities related to crop genetic diversity in Bohol in 1996. Farmer field schools and various training and capacity-building sessions in farmer-led participatory plant breeding were main components of the project, in which farmers from various communities participated. It emerged that there were not many traditional rice varieties left in the area, due to the introduction of government-released varieties which were offered with attractive incentives. Nevertheless, production had remained low, and many farmers were thus looking for better solutions. This is where SEARICE came in, offering capacity building in organic production methods and farmer-led participatory plant breeding aimed at rice varieties that would be more readily adaptable to environmental conditions in the area and the nutrition needs of the farming communities. Selection breeding and cross-breeding of traditional and modern varieties were applied. Some farmers were particularly successful in these activities, and today their varieties are used by many farmers in Bohol. Some of these varieties are served in restaurants in Tagbilaran, the capital of Bohol.

A special feature of the Bohol project is the collaboration between SEARICE and the Central Visayas State College of Agriculture, Forestry and Technology (CVSCAFT, now Bohol Island State University) in Bilar Municipality of Bohol. This started with a joint participatory rural appraisal in the village of Zamora in 1999. The assessment showed that yields were low due to pests, diseases and crop varieties which were not adapted to the particular requirements of that area. These problems were taken as points of departure for the collaboration, in agreement with the farmers. When farmers were asked to list the rice varieties they used, it became clear that they were using only seven varieties. This reflects the situation in the area where most of the old rice varieties were already lost, and introduced varieties together with the remaining ones were not performing well.

First, SEARICE and CVSCAFT recommended that the farmers switch to organic farming, arguing that this would require less investment while at the same time it would be more environmentally friendly and would provide better nutrition. The farmers involved agreed, and in the initial phase of this initiative the focus was on soil fertility management and organic pest management. Later, genetic diversity became central – in particular, the conservation, development and sustainable use of crop genetic resources. CVSCAFT was also involved in other SEARICE-initiated activities in Bohol.

From early on, the two institutions worked together in giving courses on the management of crop genetic diversity for college students. Then they decided to establish a small gene bank with a base collection kept in a freezer and an active collection kept at room temperature. The gene bank, which focuses mainly on rice, was set up on the campus premises and opened in 2001. Samples of all farmers' varieties and varieties bred by farmers through the CBDC-BUCAP project in Bohol are deposited at the gene bank – based on an agreement with the Farmers' Consultative Council, which has representatives from all the farmers' associations involved in that project.

The varieties are all freely available to farmers. An inventory conducted in January 2008 showed there were 130 varieties in the base collection and 123 varieties in the active collection: all of them registered and characterised. CVSCAFT has provided sites for demonstration trials carried out jointly with farmers and SEARICE staff. On average, about 50 farmers visit the gene bank each month and approximately five farmers per month request for seeds. They get only a small amount of seeds of each variety that they choose for the first time to test in their fields. Then, once they have decided which variety to grow, the gene bank will provide them with a somewhat larger amount of seed, depending on availability. Visiting farmers may also use the gene bank to get an overview of accessible varieties and then directly contact the farmers who maintain the varieties. The gene bank has become an important means of providing access to genetic diversity for farmers in the area and has also been instrumental for the Community Registry in focus in this story.

Why a community registry? Philippine legislation on plant variety protection

In June 2002, Republic Act No. 9168 – An Act to Provide Protection to New Plant Varieties, Establishing a National Plant Variety Protection Board and for Other Purposes – normally referred to as the Plant Variety Protection Act (PVP Act) was enacted. Whereas this Act was aimed at improving the availability of much-needed high-quality seed in the country, it created new challenges for farmers engaged in the conservation and development of crop genetic diversity. Thus, the legislation brought worry to farmers in the Philippines.

The history of the PVP legislation process is highly interesting. It is the tale of how an initiative for a genuine Philippine *sui generis* system (a system of its own kind) of plant variety protection became transformed into an attempt to make the Philippines a member of the Union for the Protection of New Varieties of Plants (UPOV) under the 1991 Act; the newest and strictest version of the UPOV Convention (see Chapter 3). It is also the story of how the USA, through USAID, exerted crucial influence on that process, through an American consultancy firm that established satellite offices in various government ministries in the Philippines. The result was an act of legislation that closely resembled UPOV '91, but due to hard lobbying work from many sides it also had some central differences. The full story is documented in Andersen (2008), pp. 308–35; see also pp. 242–5.

The PVP Act provides plant breeders with intellectual property rights over new varieties of plants as an incentive to investment in the development of new plant varieties. The holders of plant breeders' rights have exclusive rights to authorise the production and reproduction, conditioning for the purpose of propagation, offering for sale, selling or marketing, exporting, importing and stocking for any of the mentioned purposes (Section 36). These rights also extend to harvested material (Section 38), to varieties that are not clearly distinct from the protected variety and to essentially derived varieties (Section 39).

There are however, exemptions for small-scale farmers – this is one point where the Philippine legislation differs from UPOV '91. Farmers' traditions of saving, using, exchanging, sharing and selling their farm produce are termed a *right* in the PVP Act and are exempted from the rights of the plant breeders, provided that a sale is not for the purpose of reproduction under a commercial marketing agreement. The exemption also extends to the exchange and sale of seeds among and between farmers, provided that this is done for reproduction and replanting on their own land. Even though these provisions have been heavily criticised by NGOs and some farmer organisations in the Philippines for limiting Farmers' Rights, they do provide some legal space for small-scale farmers to continue their practices as long as the provisions are interpreted in line with the needs of these farmers (Andersen 2008:

pp. 242–5). To the knowledge of the authors of this chapter, this has been the case so far.

Thus, despite the heavy criticism, these provisions do not actually constitute the key challenges with regard to Farmers' Rights in the Philippines as concerns the PVP Act. The main challenges have to do with another aspect: how to establish *prior art*, i.e. the knowledge of already existing plant varieties. This is important to ensure that a plant variety for which a certificate of plant variety protection is sought is really new. A plant variety is deemed 'new' in this context if it has been bred, or discovered and developed, by the applicant of the PVP certificate and has not been sold, offered for sale, or disposed of for more than one year in the Philippines. In practice, this means that a breeder can 'discover' and further develop a farmers' variety and be granted intellectual property rights. It is therefore not difficult to understand the worries of farmers, farmers' organisations and NGOs.

In order to establish *prior art*, a Registrar is to maintain a database of existing plant varieties, collected from foreign and local databases (Section 74D). Farming communities and *bona fide* farmers are to establish community inventories (Section 72), which are then to be included in the database, according to the PVP Act. Furthermore, all certificates of plant variety protection are to be published in the Plant Variety Gazette, and objections must be filed within a period to be prescribed by the Plant Variety Protection Board from the date of certificate publication.

In short, this means that farmers are made responsible for registering their own varieties. Further, that the burden of proof rests with farmers and farmers' organisations in cases where a breeder applies for protection of a plant variety which is essentially derived from, or identical to, a farmers' variety (Andersen, 2008: pp. 242–5). To avoid misappropriation of the varieties conserved and developed in Bohol, SEARICE proposed that community registries be established and maintained in the areas where the organisation had been active. This was the start of the Community Registry project.

Organising the Community Registry in Bohol

In 2002, after the adoption of the PVP Act, SEARICE visited the Campagao Farmers' Production and Research Association (CFPRA) in Bilar, Bohol. CFPRA is a partner of SEARICE and was already then renowned in Bohol for its achievements in farmers' plant breeding. In particular, the farmer breeder Mr Cisenio Salces was famous for the red rice varieties he had bred and which became popular among farmers in many areas of Bohol. SEARICE told the members of CFPRA about the new PVP Act and consulted with the farmers about how to respond to this new situation. Mr Salces, at that time President of CFPRA, was given the opportunity to attend a regional consultation on the PVP Act conducted by the Department of Agriculture as well as a forum on Farmers' Rights conducted at the University of the Philippines in Diliman, Quezon City. He came home deeply concerned about the varieties that had been bred in the community, fearing that they could become subject to misappropriation. Also, he feared that the new Act would constitute a threat to the free availability of crop genetic

Figure 7.2 Rice fields in the Campagao community in Bilar, Bohol
   Source: Regine Andersen

resources, which was so important to ensure diversity and to the breeding work of the association.

On this basis, the members of CFPRA started to explore how to safeguard the plant varieties they maintained and had bred against misappropriation to ensure that these would remain freely available to farmers in Bohol and elsewhere as part of the public domain. With legal advice from SEARICE, they decided to establish a community registry of the rice varieties that were conserved, developed and used locally, and to declare that these varieties would be free from intellectual property rights and publicly available for farmers. The members of CFPRA chose an affidavit (a sworn statement) as the legal instrument for this declaration. A joint affidavit listing 11 rice varieties and declaring these as public domain was signed by the 27 members of CFPRA in 2003.

Furthermore, CFPRA filed a resolution before the Community Council of Campagao, recognising and supporting the Community Registry. This resolution was approved and passed. This step was important for two reasons: first, CFPRA had the backing of the village; second, the village was informed about CFPRA's activities and had been assured that the varieties listed in the Community Registry would be freely available to all.

The 11 rice varieties were deposited at the gene bank maintained by CVSCAFT, where they were also characterised according to the requirements of the gene bank.

Thereby *prior art* was established – which was the key measure under the PVP Act and necessary in order to avoid misappropriation.

The CFPRA Community Registry provided a model for other farmers' associations in the area. Experiences were shared at farmers' field days, meetings and seminars. Even if the main features were the same across communities, the various initiatives evolved somewhat differently, as we shall see.

The Zamora Organic Farmer-Researcher Association (ZOFRA) decided to seek legal recognition of their Community Registry through a Memorandum of Agreement which they filed before the Zamora Community Council. At first, the Council was somewhat reluctant to approve such a memorandum. When CVSCAFT and SEARICE intervened, however, the Council became supportive. The Memorandum of Agreement was adopted in June 2004: it supports ZOFRA's initiative to protect varieties used among local farmers – formal varieties, traditional varieties and farmer-bred varieties. It also provides that the Zamora Community Council is to help farmers in their efforts to conserve, develop and utilise plant genetic resources.

The Farmers' Association for Community Development – Riverside (KMKK) registered 11 farmer-bred varieties, four traditional varieties and five formally released varieties in 2003. Nevertheless, the draft resolution prepared by KMKK could not be passed in the Community Council, due to opposition from some council members. They believed that the Community Registry provided the KMKK with *ownership* of the listed varieties, including traditional varieties commonly shared by members and non-members alike. KMKK did not manage to convince the opposition that this was not the case and that the Community Registry was in fact meant to ensure the opposite: that these varieties would remain in the public domain.

Also for the Consumbol Organic Farmers' Association (COFA), establishing a community registry was not that simple. It turned out that the Community Council was not united and would not be willing to support a resolution. Therefore, COFA did not try to formulate a resolution to support their registry in the first place.

However, it proved possible to resolve this somewhat difficult situation for KMKK and COFA alike by moving the case up to the municipal level and filing a resolution with the Municipality Council of Bilar, Bohol. This Resolution No. 81, Series of 2004, 'A resolution recognizing the rice community registry of CFPRA, COFA, ZOFRA and KMKK, whose location is based in the Municipality of Bilar, Bohol' was adopted by the Municipality Council of Bilar, Bohol on 4 June 2004. It states that the Municipality Council finds it necessary to provide legislative recognition of the Rice Community Registry of the *barangay*s of Campagao, Cansumbol, Zamora and Riverside in order to forestall misappropriation and unfair monopolisation of varieties and resolves that the Municipality Council of Bilar, Bohol, recognises the Rice Community Registry.

Later on two farmers' associations joined the Community Registry. The Poblacion Vieja Sustainable Farmers' Association (PVSFA) started working on their registry in 2003 and produced a comprehensive resolution text with annexes on variety characterisation, including related traditional knowledge and a diversity map. It covers 11 farmer-bred varieties, five traditional varieties and six formally released varieties. The affidavit has 23 signatories, and the resolution was approved in 2008 by the Municipality Council of Batuan, Bohol, thereby recognising the Community Rice Registry of PVSFA covering the Municipality of Batuan, Bohol.

The Malitbog Sustainable Farmers' Association (MASFA) joined the Community Registry in 2005, when the Community Council of Malitbog approved the filed resolution, thereby including 13 varieties in the registry. The resolution states that the Community Council of Malitbog firmly believes that the programme is very much needed by the local farmers 'for the protection of local seeds and local knowledge from misappropriation and unfair monopolization (PVP Act of 2002)'. Moreover, it provides that the Community Council will help MASFA in carrying out its rice variety inventory every cropping season.

As we see, the six farmer associations of Bohol involved in the Community Registry have faced different challenges but have all succeeded in registering the rice varieties they maintain and breed. The main elements in this process have been the declaration or affidavit with an inventory of the rice varieties that are registered, followed by legal recognition by the local authorities in the form of a resolution or a memorandum of agreement and the deposition of accessions of the registered rice varieties in the local gene bank. Thereby *prior art* has been established as per the PVP Act, and the registered varieties cannot be misappropriated as such. How has this affected the farmers and their work to promote crop genetic diversity?

Impact of the Community Registry

Copies of the lists of registered varieties are on file in the SEARICE office in Tagbilaran, the capital of Bohol. Here, SEARICE staff can follow up and keep track of the PVP certificates announced in the Plant Variety Gazette and compare them against the varieties listed in the Community Registry. As yet there have been no attempts of misappropriation of the varieties maintained and bred by farmers in Bohol. Thus, the farmers involved in the Community Registry have not experienced how the registry system could help them in actual cases. However, they feel that it provides them with security that the varieties they maintain and breed will not be subject to misappropriation.

Now we must ask: was it worthwhile to invest so much effort in a community registry, even if the PVP Act provides for such measures and when the risk of misappropriation would probably be quite low? When the authors of this chapter visited the various farmers' associations we discussed this question with them.

The farmers told us that they had already been actively engaged in the development of crop genetic diversity before the Community Registry was established and that they felt that the registry was actually not required to ensure access to seed. After all, there had been access also before the Community Registry. The sole reason for the Community Registry was to provide protection against possible misappropriation of local rice varieties. Nevertheless, this was an important aspect, they stressed, as it enabled them to share their seeds freely without having to worry about possible misuse. And in that way, the Community Registry may have been instrumental in maintaining the traditional ways of sharing seeds also after the PVP Act.

Some farmers added that they believed the Community Registry to have been instrumental in maintaining varieties that might otherwise have been lost. The registry put in place a system that makes it possible to keep track of the varieties at hand, and to ensure that they are maintained and conserved.

Furthermore, the farmers told us that the community registry process had been important as regards awareness-raising on seed issues. Members of the various farmers' associations had gained greater understanding of the political aspects of plant breeding. They also felt that the process had brought them closer to each other as a group and that this had improved their cooperation, exchange of seeds and knowledge as well as the sharing of technical skills. Moreover, recognition through the Community Council and the Municipality Council had strengthened their self-esteem. As such, the process had led to more empowerment among the farmers. All of these farmers' associations are devoted to organic farming and farmer-led partic-ipatory plant breeding. The community registry process provided a 'glue' between the members of each association that helped in strengthening this work.

The Community Registry must be seen as one of several activities that are closely interrelated. The main purposes of these six farmers' associations relate to organic farming and farmer-led participatory plant breeding. In this way, they contribute to protecting the environment and ensuring access to safe and nutritious food. The farmers emphasised that their breeding efforts had provided greater food security as well as food that was both filling and tasty. They also saved money by not having to buy expensive farming input; moreover, they could sell their produce at a higher price. The farmers realised that they had succeeded in breeding good varieties and developing organic agriculture adapted to their needs and the specific characteristics of their environment. They received recognition and felt pride for these achievements: it was good to be self-sufficient in seeds, to have greater choice in rice varieties and even to be in a position to offer good seeds to other farmers. The popularity of the farmer-bred varieties in Bohol was seen as an important measure of the success of their efforts. And they felt that the Community Registry had contributed to this development.

Ms Ruperta D. Mangayay, farmer breeder and President of ZOFRA, formulated it this way:

> The main benefit of the Community Registry is protection from misappro-priation, and to enable our children and grandchildren to also have access to a diversity of good seed. Therefore the Community Registry is important. It goes hand in hand with participatory plant breeding and organic farming which provides us with good nutrition and higher income, especially because inorganic inputs are so expensive. Each of us maintains our diversity at the household level. With the Community Registry we feel secure when doing cross- and selection-breeding.

Challenges

A challenge mentioned by several farmers is updating the lists of registered varieties. Usually new inventories were to be made each year – at least, that was the objective. And so instead of updating the master lists, there were new lists each year covering the new varieties that were to be added to the registry. This made it difficult to get a full overview of the totality of rice varieties available to farmers and to monitor potential misappropriation. Ideally, there should be a rolling master list to be updated each year.

Another challenge is the availability of the lists. Each community would retain one copy of their own list, and in addition the SEARICE office in Tagbilaran would have one copy of each list, in order to monitor potential misappropriation as compared to the PVP certificates announced in the Plant Variety Gazette. For the original purpose of the Community Registry, that would be sufficient. However, for the Community Registry to increase the access to genetic diversity, wider distribution of the lists will be needed; for example, all the involved farmer associations could share their lists, and it would also be useful to deposit a list at the gene bank at CVSCAFT. Consideration could be given to distributing the lists even more widely, perhaps together with the affidavits and the local government resolutions – for example to the respective community councils as well as to the local government units responsible for agriculture in the various municipalities. This could also help to reduce the suspicions expressed by some local politicians that the farmer associations would seek ownership of locally used varieties and perhaps pave the way for a more positive attitude towards the work of the farmers' associations.

Even though the members of the six farming associations feel that the Community Registry process has strengthened their groups and work, they have all noted a decrease in membership. An immediate reason may be the way information is gathered for the Community Registry. Each farmer has to provide information on size of farming land and yield. This is seen as sensitive private information that could affect other matters, like the calculation of tax, so farmers are hesitant to provide such information. As this was a problem mentioned by many of the farmers consulted for this chapter, there is reason to believe that it constitutes a main hurdle to wider membership. One simple solution could be to separate such information and perhaps even not collect it, as it is probably not required for the purposes of the Community Registry.

Farmers of the different associations also noted that members of the younger generation were not so interested in plant breeding work, particularly in cross-breeding. This is also recognised as a hurdle to wider membership and to plant-breeding activities as such, and it worries many of the farmers we talked with. Whether this has to do with breeding work in particular or is more of a reflection of the general trend in many parts of the world that young people are attracted to urban life, is difficult to establish. Nevertheless, it represents a challenge: how can farming life and breeding work be made more attractive to young people?

Finally, the situation of the gene bank at CVSCAFT was challenging. The gene bank was dependent on financial support from SEARICE. However, SEARICE wanted the local government to recognise the importance of the gene bank for local farmers and assume financial responsibility for further operations. Thus, the financial support from SEARICE ended in 2009. Since then, CVSCAFT/Bohol Island State University maintains it with a small funding support from the Regional Field Unit of the Department of Agriculture.

An important question is how sustainable the Community Registry is. Although SEARICE has ended its financial support of the project, the farmers continue registering their varieties in the Community Registry and ask SEARICE for advice and guidance when necessary. SEARICE is currently working together with the local government unit to build their capacity to support the farmers.[2]

Concluding remarks

In this chapter we have seen how farmers established a community registry in response to Philippine legislation on plant variety protection in order to ensure that the varieties they maintained and bred could continue to be freely available to farmers. In fact, it emerged that the risk of misappropriation was probably not so great, as no such attempts were made. That, of course, was not possible to foresee when the PVP Act was adopted. The Community Registry has provided a safety measure for the farmers, enabling them to maintain their traditional ways of seed sharing and the exchange of related knowledge. This is an important achievement and measure of success.

As a result of this, the breeding activities of the farmers in Campagao, Zamora, Riverside, Cansumbol, Poblacion Vieja and Malitbog could continue, and farmers and farming communities in other parts of Bohol have continued to benefit from this work. Rice varieties that were considered better adapted to local conditions and more nutritious, filling and tastier have found their way from the hands of a few farmer breeders to many farmers and consumers in wider parts of Bohol, thereby contributing to the improvement of farming livelihoods and self-reliance in seed.

The Community Registry has been instrumental in mapping and documenting the knowledge related to the varieties that farmers maintain and breed; knowledge that might otherwise have disappeared. Moreover, the processes related to the Community Registry have strengthened recognition and self-esteem of the involved farmers and their belief in the importance of their own knowledge.

Whereas the original purpose of the Community Registry was to avoid misappropriation, it has greater potential. If the inventories of registered varieties are shared more widely, the registry could improve the access to the rice genetic diversity available in Bohol. Not least, if the registry were harmonised with the lists of the gene bank at CVSCAFT and made available there, that would greatly improve the accessibility of seed. As the gene bank is a natural meeting point of farmers from many different areas and not only the six associations referred to in this story, the gene bank would be a good place to have consolidated lists with all available information, including related knowledge. Visiting farmers could then get a compre-hensive overview of available varieties and consult with the gene-bank staff about the best choices. This could have considerable impact on the livelihoods of farmers and provide further incentives for contributing to the activities to increase the crop genetic diversity in the area.

In any case, this story shows how it is possible to protect farmers' knowledge and innovations in such a way that it remains available to all. We have also seen that farmers' knowledge is not necessarily limited to older, traditional varieties and ways of working with crop diversity but may also involve new and more modern approaches to enhancing the genetic pool. In Bohol, farmer breeders have managed to continue developing this knowledge for the benefit of farmers and farming communities around the island, through ensuring that the varieties they maintain and breed – and the related knowledge – remain freely available and accessible to all. This is a highly pertinent example of how Farmers' Rights can be realised through the protection of traditional knowledge.

Notes

1   This chapter is based on a visit by the authors to the six mentioned communities in Bohol in November 2008 and interviews with representatives of the farmer-breeder associations of each of these communities as well as with representatives of CVSCAFT, local government units and SEARICE. We would like to thank SEARICE warmly for facilitating our travel and visits.
2   This is based on e-mail correspondence with Nori Ignacio, Executive Director of SEARICE, 3 December 2012.

Part IV

Success stories from the realisation of Farmers' Rights to participate in benefit sharing

8 Participatory barley breeding in Syria[1]

Salvatore Ceccarelli, Stefania Grando and Tone Winge[2]

In Kherbet El Dieb, a village north of Aleppo and one of the 24 villages in Syria that was involved in a participatory plant breeding (PPB) project initiated by ICARDA (the International Center for Agricultural Research in Dry Areas), yields have been increasing ever since farmers started to adopt the varieties developed through the PPB programme there. PPB is one of the most common types of benefit sharing activities related to Farmers' Rights. By combining farmers' knowledge with the knowledge of professional breeders, this approach enables farmers to benefit from their contributions to the global genetic pool, for example through value added to their crops, livelihood improvements and increased income.

Fawaz Al-Abboud Al-Hassoun, one of the farmers from Kherbet El Dieb who took part in the project, is very pleased with the participatory approach and the resulting varieties. These new varieties have proven highly productive, due to their greater resistance to drought and cold, and for this reason they have been adopted by many farmers in the village. According to Mr Al-Hassoun, the varieties are highly appreciated by other farmers, and he thinks it would be excellent if breeding in general could become a participatory process among farmers, agronomists and breeders.

However, whether he will actually have his wish granted remains to be seen. Despite its initial fruitful collaboration with the General Commission for Scientific and Agricultural Research (GCSAR), the formal national research institution for breeding in Syria, the PPB project had to be scaled down because of the souring relationship with the government, and then came to a premature end due to the conditions of social unrest in the country.[3]

In this chapter, the PPB model developed in Syria will be presented, along with the lessons it offers for others. In addition, the challenges encountered by the project and the difficulties that led to its termination and what these experiences can teach others will be discussed.

Participatory research and plant breeding

Since the beginning of agriculture and until the re-discovery of Mendel's laws and the start of scientific plant breeding, farmers planted, harvested, stored and exchanged seeds, continued to modify their crops and as a result managed to feed not only themselves but also the rest of society. In doing all this they accumulated an immense store of knowledge. From the beginning of the twentieth century, however, plant breeding was gradually taken out of farmers' hands: what had previously been done by very many people in very many different places and with an emphasis on specific adaptation, was now increasingly being done by relatively few people in relatively few places and with an emphasis on wide adaptation.

This development took place without much consideration being given to the wealth of knowledge accumulated by farmers in the course of millennia. The difference between traditional knowledge and modern science was probably one of the reasons. Traditional knowledge is based on repeated observations by many people over long periods of time; modern science is based on repeated, controlled observations in a specific area (replications). Another difference is the way in which the two types of knowledge are communicated: while traditional knowledge is usually oral and is shared in informal ways, modern science is almost always communicated in a written and highly formal manner. It can therefore be difficult for scientists to elicit 'traditional knowledge' by using the communication tools of modern science. As a result, the knowledge of farmers is often ignored by conventional plant breeders. PPB represents one way to tackle this problem and also contributes to another important part of the concept of Farmers' Rights: protecting traditional knowledge related to plant genetic resources.

Recent years have seen rising interest in participatory research in general and in PPB in particular. Agricultural scientists, following the early work of Rhoades and Booth (1982), have become more aware of how farmers' participation in technology and variety development can increase the probability of success for the end results. This heightened interest in PPB can be explained partly by the impact of agricultural research, including plant breeding, on many farmers in developing countries, which has fallen below expectations, especially for those who farm in marginal environments (see for example Bellon, 2006). In addition, one should bear in mind that a staggering 925 million people were estimated to be undernourished in 2010 (FAO 2010a).

The limited impact of most agricultural research in marginal areas may well be due, at least in part, to the fact that the research agenda is largely determined unilaterally by scientists, and not discussed with farmers. Moreover, agricultural research tends to be organised according to disciplines or specific commodities and seldom reflects the real integration often operating at farm level. In addition, there is an imbalance between the large number of technologies generated by the agricultural sciences and the relatively small number actually adopted and used by farmers (Ceccarelli and Grando, 2002).

With regard to plant breeding programmes, there is broad consensus that plant breeding has not been very successful in marginal environments in the developing world (Bellon, 2006) and that it still takes too long, about 10 to 15 years (Ashby, 2009), to release new varieties. Furthermore, rather few released varieties are actually adopted by these farmers; many of them opt to grow varieties which have not been officially released (Walker, 2006). Even if new varieties prove acceptable to local farmers, the seed is often either not available or too expensive. Conventional modern plant breeding has also been shown to be associated with widespread loss of biodiversity: it tends to rely on a narrow genetic basis, and the resulting varieties replace the more diverse landraces in the fields (Wolfe, 2000). Changing this situation is important, both to improve the livelihoods of farmers and to maintain plant genetic diversity.

Participatory research, where users are involved in the design of a new technology – and not merely in the final testing – is one way to address these problems. PPB in particular, where farmers work together with scientists and other partners (such as extension staff, seed producers, traders and NGOs) to develop new varieties, can contribute to ensuring that agricultural research is, as noted by Bellon (2006), targeted (focused on the right farmers), relevant (responding to real needs, concerns and preferences) and appropriate (able to produce varieties that can be adopted). PPB has the potential to contribute substantially to the realisation of Farmers' Rights because it not only allows farmers to benefit from their contributions to the management of plant genetic diversity but also involves them as equal partners in decision-making processes concerning the management of these resources. It can bring back more of the control of seed to the farmers themselves.

However, despite these differences, the science behind participatory and 'conventional' plant breeding is the same. The major difference is that the latter is a process where priorities, objectives and methodologies are all determined by scientists, with little participation from farmers; by contrast, PPB gives equal weight to the opinions of farmers and scientists. It is also important to distinguish between PPB and farmers' breeding: the latter refers to the various breeding activities that farmers conduct on their own, without the involvement of scientists.

There are no fixed models in PPB. For the same crop and even within the same country, different models may be required, depending on such factors as the genetic structure of the varieties and how farmers used to handle on-farm genetic diversity before they became involved in PPB. In the model employed by ICARDA (see also Ceccarelli and Grando, 2007) on a range of self-pollinating crops (barley, bread wheat, chickpea, durum wheat, faba bean and lentil) and in several countries (Algeria, Egypt, Eritrea, Ethiopia, Iran, Jordan, Syria and Yemen), the role of the scientists is to make the crosses (mostly between landraces and between improved cultivars and landraces and wild relatives), grow the first two generations on the research stations, measure features the farmers have defined as important as well as analyse and keep a safely stored electronic copy of all the data. The farmers themselves routinely evaluate and score the breeding material, decide what to select and what to discard, adopt and name varieties and produce and distribute seed of the adopted varieties. It is the details of this model, as first implemented in Syria, that will be further explained in this chapter.

The first phase of the PPB project

The barley breeding programme at ICARDA, one of the 15 Research Centres of the CGIAR Consortium, first became involved in participatory plant breeding in Syria in 1995. Because of its perceived advantages, PPB is well-suited to ICARDA's

Figure 8.1 Farmers evaluating breeding material in a village in Central Syria
        Source: Salvatore Ceccarelli

objective of improving the livelihoods of the resource-poor in dry areas by enhancing food security, alleviating poverty through research and partnerships to achieve sustainable increases in agricultural productivity and income, and ensuring the efficient and equitable use and conservation of natural resources.[4]

The GCSAR, the national research institution for breeding, was involved in the Syrian PPB project from the beginning; in fact, the barley breeder at the GCSAR co-authored one of the first papers written on PPB (see Ceccarelli *et al.*, 2000).[5] A main objective of the project was to develop a methodology for transforming top–down centralised breeding programmes into bottom–up participatory and decentralised programmes, thereby maximising the impact and ensuring continuity and sustainability. To achieve this it is very important to have the relevant national institutions on board. An additional goal was to provide a model that could be used by national programmes in other countries and for other crops.

Project partners and responsibilities

The PPB project was designed as a continuing project which involved the two ICARDA research stations, the staff of GCSAR and the Extension Directorate. At one point, several hundred farmers from 24 villages all across Syria were involved. This was possible not least due to the collaboration with GCSAR staff from various provincial research stations, as they enjoy good relations with farmers in the villages. However, towards the end of 2007 (as will be further explained), the GCSAR decided to withdraw its support. As a result, the project was reduced to nine villages. Then the project ran into further difficulties due to the spreading unrest, and in the 2011/2012 season there were no project activities in any of the villages. Most of the involved villages were located in marginal areas, frequently affected by droughts and resultant crop losses. The breeding of varieties that would thrive under these climatic conditions was therefore an important aspect of the project, and the scientific staff of GCSAR worked together with the farmers to find the best-suited varieties.

Farmers were involved in the PPB project from the very start. In the beginning, this involvement meant consultations about the overall objectives of the project. Together with the farmers it was decided that developing new and better varieties of barley should be the main priority and that this would be done in farmers' fields and with farmers' participation.

The first part of the project was of an exploratory nature, aimed at building human relationships, understanding farmers' preferences, measuring farmers' selection efficiency, developing a methodology for scoring and enhancing farmers' skills. The exploratory work included the selection of farmers and test sites and the establishment of one common experiment in nine villages and in two of ICARDA's research stations. The nine villages represented not only a range of climatic conditions, ranging from wet to dry, but also considerable variation in literacy levels, in average farm size (from about 5 hectares to about 160 hectares), in farm types in relation to the extent of crop and livestock production, in income (on-farm and off-farm) and in the importance of barley in the local farming system.

With an average annual rainfall of 174 millimetres, Kherbet El Dieb is one of the driest participating villages. It relies on sheep husbandry. As the main livestock feed, barley therefore plays a critical role to food security in the village. Barley is used solely as animal fodder (mainly for sheep) in Syria. While it might be the only crop choice in dry areas, it can also be a rain-fed crop as part of more complex farming systems together with wheat, lentils, chickpeas and summer crops. Farmers with their own herds of sheep will use the barley they grow as feed and sell any surplus; farmers with no herds of their own will sell their entire barley harvest (both grain and straw) on the market.

The two research stations, Tel Hadya and Breda, represent two distinct production environments. Tel Hadya, with an average annual rainfall of 335 millimetres, has a typical high-input favourable environment, where barley can be grown together with a wide range of other crops. Breda, by contrast, with an average annual rainfall of only 269 millimetres, has a typical low-input, high-risk environment, where barley is the most common rain-fed crop and there is a limited range of other rain-fed crops and cropping systems.

The initial barley experiment lasted three cropping seasons (1996/97, 1997/98 and 1998/99). It included 200 new barley types that represented a wide range of variability on many characteristics such as plant height, flowering and maturity date, leaf colour, row type (two rows, six rows), seed colour (white, black, grey), stem diameter and associated lodging resistance and straw palatability. Because barley is used exclusively as animal fodder in Syria, straw palatability is a feature of high value to the farmers. In addition, eight farm cultivars from eight of the nine host farmers were also included. The 208 barley types could be sorted into several different categories. They came either from modern germplasm (100) or landraces (108), were fixed lines (100) or segregating populations (108), had two rows (158) or six rows (50), and had either white seed (161), black seed (28), or mixed seed colour (19). Both before and after planting, the agronomic management of the trials was left to the host farmer. Trials were conducted under rain-fed conditions in the farmers' fields and at the research stations, because that is the most common condition for barley growing in the area.

Each of the participating farmers was given a field book in which to record daily rainfall and evaluative plot observations. Most farmers preferred a numerical scale as their quantitative scoring method. Some used qualitative scoring such as ticking or classifying the plots as 'bad', 'medium', 'good', 'very good' and 'excellent'. Eventually, the farmers ended up using a mix of quantitative scores for some features and qualitative descriptors for others. The farmers used these scores at the time of the final selection to assign the final score. Farmers usually needed no assistance with the scoring, but where illiteracy was prevalent, farmers got help in writing their scores from other farmers or the scientists.

The selection process

Various types of selection procedures were performed as part of the project. Centralised non-participatory selection was conducted by the breeder at the research station, in this case the barley breeder of GCSAR, while centralised participatory selection was conducted by the farmers at the research station. The decentralised parts of the process were either non-participatory, with selection being conducted by the GCSAR breeder in the farmers' fields or participatory, with selection conducted by farmers in their own fields (Ceccarelli *et al.*, 2000).

The first selection was undertaken in May 1997. This was done independently by the various participants, so the breeders did not know what the farmers had selected and vice versa. The selected entries were classified on the basis of who selected them and where they were selected. For each of the nine farmers' fields, this resulted in the following six groups: entries selected by each farmer in his field; entries selected by each farmer at Tel Hadya research station; entries selected by each farmer at Breda research station; entries selected by the breeders in each of the farmer's fields; entries selected by the breeder at Tel Hadya research station; and entries selected by the breeder at Breda research station. The first four groups of entries were specific to each of the nine farmers' fields, although several entries were commonly selected in more than one farmers' field. With the selected entries, a specific trial was prepared for each of the nine farmers' fields and care was taken to avoid duplications. The entries from the two last groups were common to all trials.

In the cropping season of 1997/98, local landraces and improved varieties were used as systematic checks. These checks were chosen by the farmers. One farmer, Abdu Sheiko, from the area near Al Bab (a large village 60 kilometres northeast of

Aleppo) had introduced a forage legume crop in the rotation. The trial was therefore planted twice: once after the barley and once after the legume. All ten trials were also planted at the two research stations, using the same layout as in the farmers' fields. The total number of entries tested in 1998 was 1,348, of which 196 were genetically different as a result of the wide diversity in selection criteria used in 1997. The process of evaluation and selection conducted in 1997 was repeated in 1998 on the lines selected the first year, and again in 1999 on the lines selected in 1998.

Initial results

The results of the first three years supported the hypothesis that farmers are able to handle large populations of entries; to make various observations during the cropping season; and to develop their own scoring methods. It was also demonstrated that farmers select for specific adaptive traits and that in some cases, such as modern germplasm versus landraces, selection is driven mainly by environmental effects (Ceccarelli *et al.*, 2000). The testing also showed greater diversity among farmers' selections in their own fields than among farmers' selections at the research stations and among breeders' selections irrespective of where the selection was conducted. Interestingly, the selection criteria used by the farmers proved to be nearly the same as those used by the breeders. In addition, in their own fields farmers proved slightly more efficient than the breeders in identifying the highest-yielding lines. The breeders were more efficient than the farmers in selecting at the research station located in an area with high rainfall but less efficient than the farmers at the research station located in an area characterised by low rainfall. These findings constitute a strong argument for farmer participation.

The first phase of the participatory barley breeding project in Syria led to heightened awareness among the farmers as to what plant breeding is and what it can offer. This was evident from the number and type of questions raised by the farmers throughout the process. Requests to extend the project to other crops besides barley showed that substantial interest had been created in participatory plant breeding. In addition, the findings regarding farmers' efficiency in selection allowed the approach to be exported to other countries (Egypt, Eritrea, Jordan, Morocco, Tunisia and Yemen) after scientists from these countries visited Syria during the first project phase.

By showing that farmers can handle a large number of lines/populations, the first phase of the PPB project also discredited the belief of some, that 'farmers are simple-minded people that are not able to handle more than 20 to 30 varieties at a time.' This was essential for moving from the linear process used in the first phase to a cyclic process and a truly participatory programme.

The second phase of the PPB project

The results from the three-year experiment indicated that there was much to gain and nothing to lose from implementing a decentralised participatory plant breeding programme. A second phase was therefore started, involving implementation of a proper participatory programme with the typical cyclic aspect of a breeding programme. This meant clarifying with the farmers that the project would not be short-term but if possible an on-going, evolving activity. The farmers were interested in participating and so the project could continue.

One important feature of this second phase of the project was that the ICARDA research stations were given a different role: now they were to be used only for seed multiplication. The number of villages taking part in the project increased from nine to 11 in 2003 and to 24 in 2005, and as a result more and more farmers became directly involved. This increase was possible because of the full support given by the Minister of Agriculture at the time. The minister was the first to propose that a workshop for farmers, scientists and ministry officials should be organised. That workshop was held in Hama in central Syria in February 2003 and led to the commitment of GCSAR and extension staff to collaborate fully in running the trials, note-taking, interacting with farmers and data analysis (Ceccarelli *et al.*, 2003). This collaboration continued until late 2007.

In addition, village-based seed production was initiated in some villages. The details of the second phase (the number of lines to be tested, plot size, type of germplasm, selection criteria and issues related to seed production) were discussed in meetings with farmers held in each of the participating villages. These led to the PPB model that came to be used in the project.

During the second phase, the number of farmers directly involved in the particpatory programme varied from five to ten per village at the time of selection and from 10 to 15 per village at the time of data discussion. As a result, between 200 and 400 farmers were directly involved in two of the most important decision-making events each cropping season. Moreover, in some villages as many as 60 farmers bought seeds of the varieties selected through the PPB project. On average more than 1,000 farmers benefited each cycle.

Several farmers also started to produce seed from the selected varieties. This seed was usually either provided for free or at a slightly higher price than the seed of little known 'improved' varieties. Because they were buying seed of a variety they had seen grown in the field by another farmer with agronomic practices similar to their own, the farmers were sometimes willing to pay more than what they would for other varieties.

Testing and selecting

The testing process involved three stages: initial yield trials, advanced trials and elite trials. The initial yield trials in Syria involved 165 lines. When there was considerable diversity in the crop, and when farmers in different villages had different preferences, the initial trials planted in different villages involved different lines. The total number of lines tested across all villages could therefore be fairly large: in the case of Syria it was more than 400 genetically different lines. Since there was only one initial trial per village, great care was taken in deciding which farmer and field to choose, in discussions with the farmers. If a choice proved unfortunate, for example an atypical farmer using agronomic practices different from most other farmers from the same village, the selections made from the initial trial might not be well-suited to the rest of the village.

The advanced and elite trials, which consisted of the lines selected during the initial trials and the advanced trials of the previous year, were replicated trials involving two replications. Data from the trials were spatially analysed by the use of a statistical programme (see Ceccarelli *et al.*, 2003). This analysis produced the best linear unbiased predictors of genotypic values and several variables including heritability. This means that these PPB trials generated data of the same quantity and quality as the data generated by the Multi-Environment Trials in a conventional modern breeding programme. In addition, these trials provide information on farmers' preferences usually not available with conventional trials. Due to the solidity of the data, the resulting varieties qualify for official variety release. In several countries, including many in the developing world, this is a prerequisite for commercial seed production.

One key aspect of this PPB model is that once it is fully implemented, the lines selected are used as parents in a new cycle of recombination and selection, just as in a conventional breeding programme. The difference is that these lines are selected by the farmers themselves and may vary from location to location. This cyclic aspect has an enormously empowering effect, as the farmers see that the breeder uses their selections to produce the next generation, and they feel that their choices are valued by the breeder. A strong sense of ownership is created.

Particular care was taken to design a scientifically robust model, for two reasons. First, in this way the farmers could be provided with scientifically correct information to base their decisions on – the same type of information a breeder usually has. Second, participatory plant breeding programmes are often criticised, and sometimes rightly so, for not using rigorous experimental design and statistical analysis, so it was important to use a model that would protect the programme against such criticisms. A continuous search to improve the methodology of PPB, followed by the adoption of new experimental designs and optimised randomisation, continued till the very end of the project (see also Ceccarelli, 2012a).

As a result of the decentralised selection process and farmer participation, the PPB project led to increased crop biodiversity. The number of different entries at the end of a breeding cycle conducted in farmers' fields was higher than the number of lines which the national Syrian breeding programme tests at the beginning of its on-farm testing, which usually ends with one or two recommended varieties across the country. Many more varieties are therefore adopted as a result of the PPB programme. In the period 2006 to 2011, 93 varieties were 'potentially adopted' by farmers (a variety is considered 'potentially adopted' when farmers have named it, begun cultivating it independently and started informal seed multiplication and seed exchange with their neighbours). Compared with the less than ten varieties officially released by the Syrian Ministry of Agriculture in the last 35 years, out of which only one has been adopted by farmers (and only to a marginal extent and in the higher rainfall areas), these numbers testify to the success of the PBB project.

Farmer initiated changes

As the PPB project progressed, farmers also contributed suggestions for modifying the methodology. One farmer suggested a new way to organise the selection process, while another farmer brought up the use of mixtures.

In the beginning, visual selection in the field was organised, as requested by the farmers, on a given day near to harvest time. On that day the farmers would gather at a meeting point, a short explanation would be given to any newcomers and each farmer would be given a separate score sheet for each trial. Each farmer then scored each plot, which could take up to half a day. At the end the scientists would collect the score sheet to computerise the scores. Interested visitors would often be taken to these gatherings where they could interact with farmers.

In 2005, Majid Awad, a farmer from Bylounan in Raqqa province, one of the driest villages in the project, declared that he was not happy with this procedure. He complained that he could not concentrate properly on the scoring; a process he regarded as central to future selection. This problem, he felt, came because he had to score all the trials on one day as well as frequently be asked questions by visitors and other farmers, and visitors would sometimes walk in front of him as he was making up his mind with regard to the score. He also pointed out that even though the day of the selection was decided in consultation with the farmers, other last-minute commitments could prevent a farmer from attending, causing him to lose the opportunity to take part in the selection.

These concerns led Mr Awad to suggest that the score sheet should be distributed to all interested farmers well in advance, enabling them to decide for themselves when to do the selection. In that way, they could spend as much time as they needed, even repeating the procedure if a climatic event changed the growing conditions. In fact, that happened one year: after the original selection had been completed, there was a heat wave that the various lines reacted differently to and so the farmers decided to repeat the scoring process.

This new way of organising the scoring originally suggested by one particular farmer in one location eventually spread to the other villages, although most farmers continued to set aside one day to discuss the results of the trials with the scientists.

Another modification of the methodology initiated by a farmer was related to the use of mixtures. Since farmers in Syria generally dislike heterogeneous plots, it came as a surprise to the ICARDA scientists to learn that Abdu Sheiko had decided to mix two very different barley varieties. One of these was a two-row, lodging-susceptible but drought-resistant variety, and the other a six-row, lodging-resistant variety that produced high yields in years with heavy rainfall. When asked why he did this, he replied that he had learned about the characteristics of the two varieties by conducting PPB trials and taking notes; and because he at the time of planting did not take drought-resistance into consideration, he thought that mixing these two varieties could be a good strategy for stabilising the yields.

As with Mr Awad's suggestion, other farmers were told about Abdu Sheiko's mixtures, and farmers in the other 24 villages started mixing their leftover seed after samples had been taken to measure the yield. Farmers later reported that mixtures produced better yields than all the other varieties.

These two examples show that farmers take PPB projects very seriously and have ideas about how they can be conducted. For such projects to be truly participatory and to succeed, it is essential to take the experiences of farmers into account and incorporate their suggestions. The degree to which experiences and knowledge spread from farmer to farmer and village to village also shows how farmers learn from each other and experiment with new methods that they think might be beneficial. This willingness to learn and participate actively, as well as the informal systems of information dissemination, means that PPB projects have the potential to meet the needs of farmers and also have an impact beyond those directly involved. Such projects are therefore well-suited for promoting the realisation of Farmers' Rights through equitable benefit sharing. Moreover, they empower farmers to take part in decision-making processes with regard to the maintenance of crop genetic resources.

Participation of women

When it emerged that women farmers in Syria were interested in PPB, but were either not informed about the possibility or simply assumed they could not participate, a female researcher from ICARDA started to promote the integration of Syrian women farmers into the PPB project by combining gender analysis with action research (see also Galié, 2012). Participatory fieldwork was conducted to understand gender-based differences in agronomic management, crop preferences and needs. In addition, multi-criteria mapping was used to discuss women's expectations of the programme, their views on the validity of the current PPB process and their suggestions for improvement.

As a result, PPB activities were organised in ways that facilitated the involvement of women farmers. This was done by coordinating the events directly with the women as well as collaborating with local institutions and by creating women-only spaces. The project sought to respect local sensitivities, particularly with regard to the participation of young female farmers in public events, and to create arenas for discussion where it was easier for women to interact with male strangers.

ICARDA also felt it was important to create opportunities for women, men and ICARDA staff to collaborate, for example by facilitating mixed meetings and creating opportunities to share common concerns, and to find and implement solutions to problems. The PPB activities were evaluated together with the farmers to deal with problems encountered from a gender perspective.

Gender issues were also taken into account in knowledge sharing. Because Syrian women tend to be less literate than men and have less access to technology, the reports from the PPB project were produced in digital and hard copy and included visual and oral material.

In addition to these changes in approach and methodology, the PPB project started to enlarge its research portfolio beyond barley to reflect women's priority crops – such as chickpea and cumin – and to include priority traits for selection suggested by women, such as hardness of the spike necessary for hand harvesting and palatability.

Figure 8.2 Women farmers evaluating breeding material in a village in the south of Syria
Source: Salvatore Ceccarelli

A key challenge in achieving gender-balanced PPB in a traditionally patriarchal country like Syria is to ensure that the participation of women farmers becomes an empowering and enriching opportunity for them, their households and communities. Then the participation of women in public spaces can be supported rather than resisted by their communities, and the benefits of the programme can be shared more equally between men and women. This is crucial to true and equitable benefit sharing, as envisaged in the Plant Treaty.

Benefits of the PPB project

The PPB project has benefited the participating villages in several ways. One especially important benefit has been the good yields from the resulting varieties. Data from what were the final years of the project, including the very dry year of 2008, show that the PPB lines outperformed both the commonly used landraces and conventionally bred modern varieties. In Kherbet El Dieb, which received 189.5 millimetres of rainfall in 2006, 206 millimetres in 2007 and only 139.0 millimetres in 2008, four PPB lines out-yielded the local black-seeded landrace grown by most farmers, by between 12.3 to 23.2 per cent. The four lines were also scored higher than the landrace by Mr Al-Hassoun and the other farmers during their visual selection. The farmers from Kherbet El Dieb estimate that in 2009 about 5,000 hectares of the cultivated land in the area was planted with varieties that had been introduced through the PPB programme four years earlier and then multiplied by the farmers. This estimate is based on the amount of seed sold and distributed and illustrates how successful the project has been in getting varieties adopted.

In Om El Amad, a village in the province of Hama with an average annual rainfall of 249 millimetres for the period 2005 to 2009 (ranging from 183 in 2008 to 328 in 2007), the two best lines out-performed the local white-seeded landrace by between 11 per cent and 19 per cent and surpassed a conventionally bred modern variety by 5 per cent to 13 per cent. In Bari Sharky, a drier village in the same province with an average rainfall during the same period of 204 millimetres (ranging from 130 millimetres in 2008 to 238 millimetres in 2005), the greatest yield increases were obtained with two lines that had resulted from crosses with the wild progenitor of barley. These lines out-performed the local landrace by about 33 per cent in terms of yield.

The selected lines have proven themselves superior not only in the most marginal and drought-affected areas. In Suran, another village in the Hama province, average annual rainfall during the period 2005 to 2009 was 277 millimetres. In three of these years it received more than 300 millimetres, whereas in the dry year of 2008 it received as little as 198 millimetres. The yield increase of the best lines – two sister lines obtained from crosses with landraces – out-performed the local landrace by 15 per cent to 25 per cent and a conventionally bred modern variety by 18 per cent to 27 per cent. Up until 2011, all these lines were grown by farmers in the four villages, and the seed was distributed to other farmers. According to Ali Turkia, from the village of Tel-Hassan Bash, all the farmers who saw how the 'Yana mixture'– a mixture of seed from the advanced, elite and extended trials in his field – grew, asked for seed for the next season. In particular, they liked the plant height and spike length. Compared to the local barley variety of this area as well as the conventionally bred Furat 2, this mixture performed very well. These findings demonstrate some of the yield increases and livelihood improvements that can be achieved through farmer participation in the breeding process.

A study of PPB and conventional barley breeding in Syria (Mustafa *et al.*, 2006) supports the hypothesis which holds that regardless of the number of new varieties released by the formal breeding sector and how big the yield gains are over local varieties, farmers in marginal environments tend not to adopt them unless they have participated in the selection. This makes PPB a particularly important tool for benefit sharing: indeed, it might be one of the only ways to ensure that these farmers benefit from scientific methods, knowledge and discoveries. In that study, analysis of the farm-level costs and benefits related to barley production indicated that the participation of farmers in breeding programmes does not necessarily entail higher production costs. Taking estimated adoption rates and yield gains as the point of departure and comparing them to investment costs, Mustafa *et al.* (2006) calculated that the cost–benefit ratio was substantially higher for PPB than for conventional barley breeding. Even assuming that only 50 per cent of the adoption rate and 50 per cent of the yield gain would be obtained for the varieties of the two programmes, the analysis showed that the cost–benefit ratio as well as the internal rate of return would be higher for PPB. This finding still held with an adoption rate of 10 per cent and 33 per cent yield gain.[6]

These findings demonstrate the importance of PPB and farmer participation and show that there is more to gain by implementing PPB than by continuing conventional plant breeding. However, it is important to ensure that all interested farmers, most of whom live in marginal areas where drought is a major risk factor, actually have access to seeds as that will determine the impact of the PPB project (Mustafa *et al.*, 2006).

One of the most important benefits the PPB project in Syria has brought to the participating farmers, as shown in Mustafa *et al.* (2006), is the positive impact on their livelihoods. When interviewed about the effects of the project, 65 per cent of the farmers said that their livelihoods and economic status had improved as a result of their participation. Most of the farmers who had not yet felt this positive impact lived in areas where the PPB project had started later and had thus not come as far (Mustafa *et al.*, 2006).

Through questionnaires and interviews Mustafa *et al.* (2006) also show that the Syrian farmers benefited from the PPB project in ways other than economically. Many of the farmers interviewed felt that by participating in the PPB project, they had learned more about barley production, agriculture in general and variety selection. This learning seems to have resulted from the interaction with scientists as well as with other farmers as almost as many (21 per cent) responded that they had gained new knowledge from the latter type of interaction as from the former (27 per cent, Mustafa *et al.,* 2006). This indicates that the project resulted in an increase in *social capital,* here defined as the ability to cooperate and share information, as well as in *human capital.*

In addition, almost all of the farmers interviewed said that even if the PPB project ended they would continue to practise what they had learned. They also thought that they would maintain the new varieties and keep looking for good varieties together with other farmers (Mustafa *et al.*, 2006). And indeed, after the project had to be terminated, all the farmers who had been in contact with formerly involved scientists from ICARDA up to January 2012 confirmed that they were still growing the varieties selected during the project period. In the autumn of 2011, one farmer was even able to sell 60 tonnes of seed from one of the PPB varieties. The fact that the farmers have chosen to keep on cultivating the PPB varieties, despite insufficient seed supplies due to the 2011 drought and the difficulties in harvesting caused by the social unrest, clearly shows the value placed on these varieties by the farmers, and the important contributions made by the PPB project.

Moreover, only a very small number of the farmers interviewed believed that those who select new varieties should keep the economic benefits resulting from seed production (etc.) for themselves: the majority held that the benefits should be shared and distributed at the community level (Mustafa *et al.*, 2006). This may indicate that the farmers view the local plant genetic resources as their common heritage, and not something only a few should benefit from. And that in turn indicates that by taking cooperation, sharing and equal distribution of the benefits as their point of departure, other benefit-sharing projects may be more in tune with the values of the local farming communities.

Legal aspects

It is commonly thought that the legislation in Syria regulating variety release and seed multiplication and distribution has functioned as an obstacle to the project on participatory barley breeding by limiting the amount of seed that can be produced and distributed; thus preventing thousands of farmers from benefiting from the project. As will be explained in the next section, the existence of such legislation was also cited when the government maintained that the PPB project was illegal. In fact, however, the sole piece of legislation existing on this subject is a Ministerial Decree from 1975 (available only in Arabic), and this document does not contain any specific restrictions regarding the movement of barley seed. This indicates that the legal situation in Syria is somewhat unclear on this issue and that the uncertainties surrounding the legality of seed distribution have been a barrier to scaling up the project. The Ministry of Agriculture and Agrarian Reform has begun drafting a Seed Law. This law seems likely to bring greater legal certainty, but depending on the restrictions it places on the rights of farmers to exchange seed it might also prove detrimental to Farmers' Rights.

Institutional challenges

It was expected that the success of the PPB programme and the fruitful collaboration with GCSAR would lead to the institutionalisation of PPB in Syria. The GCSAR is a very central institution when it comes to distribution of and access to plant varieties. In addition to being responsible for plant breeding, it is also responsible for the evaluation of candidate varieties for release, for deciding which varieties to propose for release and for seed legislation.

The national release process is conducted by the GCSAR. It involves three years of testing in on-farm field trials, at the end of which the breeder responsible proposes one or more lines for release to the national Variety Release Committee, which is chaired by the Minister of Agriculture. However, on-farm trials as practised in Syria suffer from several problems evident also in many other developing countries (see Tripp *et al.*, 1997). In Syria the main problems are as follows:

- The trials are sometimes not planted in farmers' fields: in Hassakeh province, for example, they are conducted at the Al Majarja research station, where grain yields are on average 2 tonnes higher per hectare than elsewhere in the province.
- Randomized Complete Block Design is still used as the experimental design, even though the GCSAR staff has been trained to use the more sophisticated and more accurate experimental designs used by most breeding programmes today.
- Statistical analysis is done with outdated software: this results, *inter alia*, in trials being discarded when some entries fail and no adjustments are made for spatial variation, although the GCSAR staff has been trained in the use of modern software.
- The trials are not representative: most are conducted under fallow (a rarely used form of rotation in Syria) and with the 'recommended' dose of fertilisers –

whereas in the case of a crop like barley, which is grown predominantly in marginal environments, farmers rarely use any fertiliser; moreover, the field with the most favourable conditions is often chosen, instead of the one with conditions most similar to those of local farmers' fields.
- Farmers' opinions are not included when data are collected.
- The system does not guarantee the release of disease-resistant varieties, as shown by two recently released barley varieties: Furat 2 (susceptible to scald) and Furat 7 (susceptible to powdery mildew).
- Grain yield is seen as the decisive trait even though other traits, such as plant height under drought, are important for actual adoption in dry areas.

As the PPB trials did not suffer from any of the problems listed above, the ICARDA staff had hoped that the data generated by the PPB project would be accepted by the GCSAR and that the best PPB lines could be proposed for release.

However, even though the same GCSAR staff members involved in collecting data for the variety release process had also collected all the data from the PPB trials in 2003 to 2007, the data were not examined for a possible release. The PPB methodology was not adopted by the National Breeding Programme. The closest it came was in 2007 – when despite four years of testing in farmers' fields with consolidated support from the farmers and even though all the data had been analysed in collaboration with GCSAR staff, the GCSAR proposed that the lines selected, named and *de facto* adopted by farmers should be tested *again* for three cropping seasons and that only the data from these additional three years should be used. The reason given was that relevant legislation must be respected.

This proposal was discussed with the farmers, but both the project staff and the farmers felt that such a solution was not acceptable because it would delay the entire process unnecessarily, and any resulting additional information would be of dubious scientific value. Furthermore, there was no guarantee that the farmers' best interests with regard to the varieties would be taken into consideration. The project staff tried to consult the legislation referred to in these discussions but found no existing provisions that would pose a barrier to their work.

As mentioned, in 2003 the Minister of Agriculture had been very supportive of the PPB project, which made it easy to ensure the support of the relevant institutions and their staff. However, when a new minister took over in 2005, this support started to erode, and the disease resistance and genetic stability of the lines in focus were repeatedly questioned. Another main argument used was that the project was in contravention of the law. During the closing ceremony of a workshop held in Jordan in late 2007, a representative from the Syrian Ministry of Agriculture claimed that some of the activities described by the farmers in their presentations were not compatible with the current seed legislation because the farmers had been cultivating varieties that had not been officially released and had also commercialised the seed. The Syrian farmers present reacted angrily to these accusations; one of them asked who had made laws that prevented him from growing a variety which had the potential to improve his life. A few weeks later the Minister of Agriculture sent a letter to the Director-General of ICARDA, claiming that PPB activities represented a threat to the national food security. The letter was followed by instructions to the staff of the GCSAR and the extension services to refrain from collaborating in PPB trials. This happened only a few days before the 2007 sowing for the 2007/2008 season. The project was forced to restructure, and went from covering 24 villages to a mere nine. Then, with the unrest that started in 2011 as a contributing factor, it eventually terminated its activities, at least temporarily. No PPB trials were planted in the 2011/2012 season.

The way forward – evolutionary plant breeding

For PPB to have a large impact on farmers in any country, cooperation on the part of central national institutions is essential. Expansion and institutionalisation of projects that have proven their worth is important for realising Farmers' Right in a way that can reach as many farmers as possible. As we have seen here, such an approach was initially highly successful in Syria, as the support of the government allowed the PPB project to expand its activities to a substantial number of villages. However, as became sadly evident, changes at the top can lead to sudden policy changes; and especially in countries where transparency and democratic procedures are weak, such changes can be difficult to predict, not to mention overturn. In such contexts, evolutionary–participatory plant breeding might be a useful approach, as its success is less dependent on institutional support. Evolutionary plant breeding was initiated in Syria as a continuation of the PPB project before it was shut down (see also Ceccarelli, 2012b). Since then, the breeding efforts have continued, and at least one farmer sowed one of the resulting populations in the 2011/12 season despite the difficult situation in the country.

The concept of evolutionary plant breeding was first introduced by Suneson (1956). At its centre is the use of broadly diversified germplasm, in the form of large populations, and long-term natural selection processes in the relevant areas to produce highly adapted crops. Handling these complex populations, created by mixing a large number of diverse germplasm, is in fact simple: all that is needed is to cultivate them in locations affected by either abiotic stresses (drought, high and low temperatures, salinity, soil deficiencies) or biotic stresses (diseases, insect pests) or both, and then let natural selection slowly increase the frequency of the best-adapted genotypes. With the experience and skills developed through PPB, farmers and breeders can superimpose artificial selection for traits which are important in each specific location. Evolutionary plant breeding projects can be started without being preceded by PPB projects, but the process will be smoother if the participating farmers already have experience from PPB. Different farmers may select different plants and grow the results in their own fields; this can be repeated over the years. The expectation is that the varieties derived from this evolving population will at any time be better adapted than those derived earlier.

Evolutionary plant breeding makes farmers less independent on institutions: once they receive the seed they can handle it on their own – although they are advised not to plant *all* the seed, to avoid losing the population in case of major disasters.

Farmers can actually make their own populations, by mixing all the varieties or populations available on the formal seed market. Another advantage of evolutionary plant breeding is its simplicity and enormous potential to adapt crops – any crop – to climatic changes as well as other agronomic changes which might occur in the future. In areas where PPB projects have been conducted, evolutionary plant breeding can serve as a very useful and self-sustaining follow-up that ensures that the farmers continue to benefit from maintaining crop genetic resources and from scientific developments.

Concluding remarks

The successful results of the PPB barley project in Syria have inspired other countries in the region to start participatory plant breeding of several crops. This story can also stand as an example of transfer of technology, as several PPB programmes have been initiated in European countries recently (for instance, in France, Italy and the Netherlands) based on the model developed in Syria. This indicates that this approach has great potential in a range of locations and that the lessons from Syria can continue to contribute to the realisation of Farmers' Rights also in the years to come.

A further lesson to be drawn from this story from Syria is that efforts at institutionalising PPB in existing national structures and cooperation with the relevant national institutions can be very useful and central in maximising project impact as well as for ensuring continuity and sustainability. However, as has been shown, initial support from and cooperation with the government is no guarantee for continuing support. The appointment of new individuals at the top might significantly change the working conditions for such projects, especially in countries with top–down governing structures.

To ensure project continuation and enable outreach to many farmers, it is also crucial that any barriers that seed laws might pose are examined and dealt with early in the process. Such laws are increasingly becoming the major barrier to greater impact also in other countries where this PPB model is being implemented. It is essential to consult the relevant national legislation, to see whether it places limitations on the sale and distribution of seed from unregistered varieties and under what conditions PPB-bred varieties can be registered. As the situation in Syria demonstrates, confusion as to the restrictions under existing legislation may pose a barrier to the expansion of projects that rely on being able to distribute seed.

Another important lesson to be drawn from the successful approach and results of this project, despite its termination, is if the project is to meet farmers' needs, the farmers themselves must be involved at all stages of the process. One way to ensure that farmers benefit from scientific developments is to do what this project did, giving them the opportunity to influence the development of technologies so that these can be better adapted to the farmers' specific needs, agro-ecological environments and cultural preferences. Through the project, local farmers had the opportunity to influence how financial resources for research and agricultural extension services were used. In this way, there is a greater likelihood that the resources set aside for this purpose actually benefit the intended recipients. In addition to its contribution to Farmers' Rights through benefit sharing, the PPB project in Syria also contributed to the protection of traditional knowledge relevant to plant genetic resources for food and agriculture, by using the traditional knowledge of the involved farmers. This elevates the profile of both the knowledge itself and the knowledge holders, creating incentives for continued use and development.

This project has also shown how contact with professional breeders can make farmers more aware of what science has to offer. This awareness can have an empowering effect as seen in the enhanced quality of the Syrian farmers' participation over time and how they ended up as true research partners. The farmers became involved not only in the breeding activities but also in the registration of the resulting varieties as well as in maintenance, seed multiplication and distribution, and as appropriate, commercialisation. Farmer involvement and feeling of ownership makes a project more sustainable in the long run.

This story from Syria also shows that it is possible to structure PPB in a way that makes it easier for women to participate. A gender-sensitive approach like the one implemented in this project can provide better conditions for women to take part in and benefit from the project.

The PPB project in Syria has also demonstrated how local seed systems can be strengthened, by improving the production and selection of and access to seeds. Together with the increased yields, this is a central contribution to increased food security in Kherbet El Dieb and the other villages involved in the project. The fact that the project provided some villages, such as Kherbet El Dieb, with the necessary equipment for seed production might help to encourage the continuing cultivation of the PPB varieties after the end of the project.

Despite its premature termination, the PPB project in Syria managed to bring substantial benefits to the participating villages. During the last meetings the staff had with the farmers, the farmers underlined that PPB had brought them important benefits, not least the capability to handle evolutionary populations. Looking at the experiences of this project as a whole one may conclude that, not least because farmers function as equal partners, PPB and evolutionary PPB have proven to be very useful approaches to benefit sharing and the realisation of Farmers' Rights.

Notes

1   A version of this chapter was published in Ruiz and Vernooy (eds), 2012.
2   The authors are grateful to Dr Ronnie Vernooy and Professor David A. Cleveland for reading a draft version of this chapter.
3   The unrest and conflict in Syria are usually seen as part of the 'Arab spring'. The first acts of protest took place in late January 2011, but the protest movement did not gain momentum until March that year. Demonstrations for political reform and civil rights were met with severe crack-downs by the government. Since then the conflict has escalated severely. According to the UN the Syrian government has failed to comply with almost every aspect of the peace plan from April 2012, and from May that year international pressure intensified as the violence continued.

4   For more information see the ICARDA website www.icarda.org
5   This paper won the 2000 CGIAR Chairman's Science Award for Outstanding Scientific Article.
6   For further details see www.impact.cgiar.org/assessing-benefits-and-costs-partici-patory-and-conventional-barley-breeding-programs-syria

9 Benefiting from diversity in Nepal

Tone Winge, Regine Andersen and Pratap Shrestha

If farmers around the world are to maintain crop genetic diversity in their fields, they need to see how the conservation and use of this diversity can benefit them and improve their livelihoods. If they cannot see any opportunities for benefiting from cultivating a wider diversity, for example by including traditional and local varieties, more and more farmers will probably opt to focus on only a few varieties of high economic value. In consequence, many communities may come to lose what still remains of their remaining crop diversity and related traditional knowledge, along with the valuable opportunities these resources might bring.

This chapter tells the story of what happened when farmers in the hills around Pokhara in central-western Nepal[1] became aware of the value of their crops and how their activities contribute to the realisation of Farmers' Rights. The story shows how benefit sharing can take place through the adding of value to produce, marketing efforts, knowledge sharing and cooperation.[2]

From farmers' fields to supermarkets

Ms Maina Thapa, who lives with her family on a farm in Chaur village, which is part of Begnas VDC[3] outside Pokhara, is one of the many Nepalese farmers who has benefited from the utilisation of local crop diversity. Begnas is located in the middle hills of Nepal (600–1,400 metres above sea level), with a yearly mean temperature of 20.9°C and mean annual precipitation of 3,979 millimetres (Rana *et al.*, 2011). Ms Thapa is a member of the Pratigya Cooperative, a local farmers' cooperative established in 1991 through a CARE Nepal project to organise farmers and engage them in income-generating activities, including marketing of farm products. However, the cooperative found it difficult to maintain its activities when CARE Nepal phased out its support. Then, in 1997, the Nepalese NGO LI-BIRD (Local Initiatives for Biodiversity, Research and Development) conducted an exposure visit to the area, looking for farmers' groups and institutions to work with as part of the Nepalese component of the global project 'Strengthening the Scientific Basis of In situ Conservation of Agricultural Biodiversity On-Farm' (hereinafter: the In Situ project) coordinated by the International Plant Genetic Resources Institute (IPGRI, now Bioversity International). Pratigya Cooperative expressed interest in the project and was among the groups that were selected. In this way, the cooperative was

Figure 9.1 Some of the members of the Pratigya Cooperative in front of their meeting
house
Source: Tone Winge

revived and began working in partnership with LI-BIRD, focusing on on-farm
conservation and sustainable use of agricultural biodiversity. It was also formally
registered as a cooperative under new legislation in Nepal. In 2006, the cooperative
joined Jaibik Shrot Samrakshan Abhiyan (JSSA) or 'the Biodiversity Resource
Conservation Campaign', a regional umbrella organisation which also consists of
two other nodal organisations.[4] All 14 farmers' groups in the area are now linked to
the JSSA.

Documenting and maintaining diversity

Documentation of agricultural biodiversity has been central to the work of the In
Situ project in the area and to the Cooperative's efforts. This documentation has
been carried out through the creation of community biodiversity registries.[5] In 2004,
when efforts related to community biodiversity registries in the area were scaled up
by LI-BIRD, Pratigya Cooperative took on responsibility for coordinating the
documentation of agricultural plant genetic resources (food crops, vegetables and
fruits), while the other two nodal organisations of the JSSA coordinated the documentation of forest plant species and wetland animal and plant species.

The Cooperative's work consisted, among other things, of coordinating the 14 farmer groups, each of which prepared a registry. First, these groups came together at a two-day meeting organised by the JSSA, where information was provided on biodiversity documentation. The groups then returned to their communities to work with particularly knowledgeable farmers, usually elderly men and women, in documenting agricultural plant genetic resources, their known properties and associated traditional knowledge, in a systematically organised register. Each farmer group keeps and manages the original copy of its own registry, but a registry consisting of the information from all 14 groups was also compiled by the JSSA and is kept at their headquarters. The JSSA is planning to update this registry with the help of the Pratigya Cooperative. As a result of the experience gained from this documentation process, the Cooperative has also been able to provide technical advice and support to farmers' groups in other parts of the country regarding the documentation of agricultural biodiversity.

Analysis of the documented information revealed that there were many more varieties in the area than people had thought. As the associated traditional knowledge was also collected, they became aware of the value and potential of these resources. Documenting traditional knowledge in this way contributes to its conservation and makes it available to the entire local population. In 2009, the community biodiversity registry complied by JSSA contained as many as 72 landraces of rice. Of these, 15 had originally been registered in Chaur village.

In Nepal, the extent to which traditional varieties are cultivated varies from area to area. In the hills around Pokhara for example, traditional varieties still dominate rice production, while on the plains bordering India it is modern commercial varieties that dominate (Jarvis and Hodgkin, 2008).

Heightened awareness of the existing local crop diversity can be seen as one of the main advantages of documentation of biodiversity. According to members of the Pratigya Cooperative, very few people in the village knew very much about agricultural biodiversity before the documentation process was initiated, and even the most knowledgeable individuals knew relatively little about the diversity cultivated by others. Now, however, there is a greater awareness of the rich biodiversity to be found in the area, as well as greater recognition of the value and usefulness of this diversity.

The documentation work conducted by the farmer groups and coordinated by the Pratigya Cooperative convinced the involved farmers of the importance of maintaining and sustainably using agricultural biodiversity. It also contributed to raising general awareness regarding such conservation in the area. The community biodiversity registries showed that some crop varieties were grown only on a very small-scale and only by a few people and that motivated the farmers to take action to maintain these varieties on-farm.

In particular, two measures proved useful for ensuring the continued cultivation of species and varieties that had been identified as endangered: creating a system to make sure that these varieties would be cultivated in the area, and providing opportunities for farmers to benefit from this cultivation by creating higher demand for the products through value addition and marketing.

Pratigya Cooperative has developed a system where each member of the cooperative is responsible for maintaining one or more of the varieties that has been identified as endangered. In addition to growing these varieties every year in their fields, the members also store the seeds in their own houses. If the member responsible for maintaining a certain variety chooses not to grow it a particular season, he or she must find somebody else to do so. This system of continuous cultivation ensures that the varieties evolve along with the environment and that viable seed remains available.

Access to a revolving fund provides an additional incentive to contribute to the maintenance of these varieties, as the members must assume responsibility for conserving at least one variety in order to receive a loan from the fund. The fund was established with resources from LI-BIRD but since then the Cooperative has added to it. The purpose of this fund is to help low-income groups in the community; loans are given to income-generating activities, primarily agricultural. The interest from the loans also adds to the fund.

The Pratigya Cooperative has also invested in an electric mill. This enables them to mill, package and market different varieties separately. Through such specific packaging and marketing, it is possible to increase the demand for individual varieties: thus, the mill stands as an example of how a practical measure can promote on-farm conservation through value addition. The mill has also provided the Cooperative with the opportunity to identify new varieties, when farmers bring their varieties for milling. With every newly identified variety, the cooperative takes a small portion of seed to give to one of the members for conservation.

Organising the Cooperative and the benefits of cooperation

Forty households in the village are now members of the Pratigya Cooperative, and together they hold 76 shares. Every month the members meet to review progress, discuss collection and disbursement of loans, and plan new activities. The decision-making process is consensus-based, and loans from the revolving fund are provided according to rotation-based guidelines. The Cooperative has an Executive Committee that consists of seven elected members, with a chairperson. There are five sub-committees, in charge of the management of agricultural products, marketing, mobilisation of *dalits* (the lowest-caste group), drinking water and visitor stays. Women make up the majority in these sub-committees. One male member of the Cooperative described the importance of the women in this work: 'when we take the wrong direction, the women make sure we take the right one instead.'

The members of Pratigya Cooperative feel that being part of a group has been very beneficial, and underline how everyone works together. They feel they are stronger together and can do more as a group than they can individually. Previously, they had felt unable to speak out to the same extent: now they receive more recognition and can find common solutions to their problems.

Members also underline the social dimension of collaboration and how the Cooperative has been able to remove some of the barriers between the different castes through its inclusion of families from the lower castes. As of 2012, the

Cooperative has 15 members from *dalit* families. As a result of their membership in the Cooperative the living conditions of the lower castes have improved and they have become empowered. It is now easier for those of the lower castes to speak out among the higher castes, as the Cooperative has instilled a feeling of unity in the members. The initiative has helped to decrease discrimination against poor and *dalit* families; setting an example of how social inclusion can be achieved. For its innovative work in this regard, the Cooperative has been recognised nationally and internationally, and it receives a number of visitors throughout the year. The members appreciate the recognition they now receive and are proud of the fact that, due to their efforts, they themselves have visited other areas of Nepal and have even been abroad.

They also feel that they have benefited greatly from their focus on maintaining agricultural biodiversity. The higher incomes constitute an important part of this. All in all, the members are very pleased with their cooperation and the results and claim that if the whole country had operated in the same way as them, Nepal would already be a developed country.

Improved livelihoods: Anadi and Maina Thapa's story

The story of Maina Thapa illustrates how the livelihoods of the members have improved as a result of their cooperation and activities. Because of her participation in the cooperative and her cultivation of *anadi* rice, Ms Thapa's income has improved considerably and she can now be considered a relatively affluent farmer.

Anadi is a sticky and glutinous rice variety, valued for its nutritional and medicinal properties as well as its role in traditional celebrations. The registry made the Cooperative realise that fewer and fewer farmers were growing this variety and the area under cultivation was gradually declining. Members feared that it might disappear from the area and decided to take immediate measures to ensure its continued cultivation by focusing on adding value and marketing.

In collaboration with LI-BIRD, the Pratigya Cooperative developed and carried out promotional activities for *anadi* and other products. They disseminated information on medicinal value, organised workshops and seminars, visited fairs and festivals, advertised on the radio, distributed pamphlets and ensured that their products were readily available. In fact, a majority of the urban population were already familiar with the products but were unable to find them locally. Establishing links to the urban market (explained in the next section) was therefore a key ingredient in the Cooperative's success.

As a result of these efforts, the demand for *anadi* rice started to grow and its price could increase. Now, many farmers in the area cultivate *anadi*. The Cooperative's success in this respect shows how local farmer-driven initiatives that focus on market incentives can promote on-farm *in situ* conservation of crop genetic resources.

Ms Thapa now produces an average of about 2 tonnes of *anadi* rice (some 1.17 tonnes of milled rice) a year because she has been able to access more land. She leases land from the local school, and due to the soil quality of these fields she can grow a large amount of *anadi* rice. Part of the explanation for her growing income is that the price for *anadi,* although it varies, is quite high – in 2011 it was 125 rupees per kilo, whereas for the best-quality local varieties in the Pokhara area, such as *jethobudo,* the price was around 80 rupees per kilo and for other rice types around 50 rupees per kilo. This means that even though *anadi* rice can only be grown under certain soil conditions such as heavy clay soil with high water holding capacity and fertility, and despite the relatively low yields compared to many other varieties, Ms Thapa gets a higher income from growing this variety than she would for most other rice varieties. In 2011, the 130,000 rupees she earned from the sale of 1.04 tonnes of milled rice made up the biggest part of her income, and altogether almost two thirds of her income came from activities initiated by the Cooperative. This has made her family less dependent on the pension her husband receives from the Indian army, and her increased income and independence means that she can invest the profits as she sees fit.

Her increasing success with *anadi* rice ever since she started growing this variety in 2003 as part of the Cooperative's initiative and the good income it now gives her, has enabled Ms Thapa to pay for electricity and her children's school fees. She considers herself to be much better off now than she was before the establishment of the Cooperative and her own investments in *anadi* cultivation. Previously, she relied mainly on subsistence farming and did not earn very much. After she started growing *anadi* and became involved in the Cooperative's value addition and marketing activities, her income has grown considerably. In 2006 she was even able to take up a loan to buy a tractor, which she has since managed to pay back. Trusting in her experience and confidence, the Cooperative has given Ms Thapa responsibility for coordinating the collection and processing of *anadi* rice. Ms Thapa appreciates the recognition she receives for her work and enjoys her status as a rather well-known *anadi* farmer in her area.

Selling the products

Initially, the Pratigya Cooperative did not have much success selling their products as they lacked marketing skills and experience. However, after LI-BIRD helped the members establish contact with traders in Pokhara and provided technical and financial support for packaging and labelling and initially also transport, sales slowly improved and the Cooperative started to sell their products through these traders.

However, for a while, relations with the traders were somewhat troubled as the farmers felt the traders were taking too high profits on the rice. The farmers would compare the price they got and the final retail price – without considering, for example, the costs associated with transport and any losses that might take place before the products reach the consumer. Then, when the Cooperative rented a market stall in Pokhara to sell the products themselves, they became more aware of the various costs and concluded that the traders' prices were not unreasonable. The Cooperative also found the retail business to be complex and risky and decided to approach the traders again to re-establish relations. The members of the Cooperative are now happy to sell their products to wholesale traders, as this means they themselves only have to worry about their own production costs. And as the

Figure 9.2 Maina Thapa in front of her house
Source: Tone Winge

Cooperative has come into contact with more traders, the farmers have been able to compare the competitive advantages of various traders and choose which ones to sell to.

In addition to facilitating contact between farmer groups and traders, LI-BIRD has helped the traders with the marketing of the products of the Pratigya Cooperative. Among other things, it has supported advertisements, for example on the radio, and provided software. However, the products now sell so well that the traders and middlemen need less assistance; LI-BIRD is also very conscious about not making the various parts of the value chain too dependent on them. That is why they prefer indirect support, as through advertisements, to direct support.

Janaki Agro-products, which is run and owned by Deepak Timilsina, is among the private traders LI-BIRD has supported and put in contact with farmers' groups and cooperatives. This company has a mill and a packaging machine and processes local agricultural products from the surrounding villages, including rice, finger millet, taro, beans and dry vegetables. Mr Timilsina says that *anadi* rice is quite popular and that demand for his products is increasing. However, processing can be challenging as the products must be sold within a certain time period to maintain their quality, largely because no chemical additives or preservatives are used. LI-BIRD has supported this firm in a small way, by covering the fees and transport associated with having a stall at a trade fair.

Janaki Agro-products sell their products to various retailers in Pokhara, such as Binayak and Saleways supermarkets and Namuna Gaunle shop ('the model village shop'). The majority of Namuna Gaunle shop's customers are people who have moved to the city from the villages surrounding Pokhara and are looking for the products they remember from their childhood homes. According to the owner, the store is not extremely profitable but they earn enough to get by. LI-BIRD has supported nutrient analyses that provide information for the labels and contributed to research on the durability of the products as well as providing help with advertisements and labels.

With relations established between farmers and traders and the growing demand for the farmers' local products, less assistance is required from LI-BIRD. The intention has always been to gradually scale down support to the various stakeholders as they become increasingly able to manage on their own and to interact with each other without assistance. This process is well underway with regard to the Pratigya Cooperative and its trading partners and is one of the reasons this project can be regarded as a success story.

Success despite challenges

One challenge facing the Pratigya Cooperative and other groups conducting similar activities in the area around Pokhara is that they have few younger members, and the average member age is quite high. The younger generation does not seem to be very interested, and many young people have moved to the cities. This has resulted in almost empty villages in some areas and villages consisting mainly of the elderly in other areas. In general there are few young adults left. This 'disappearance of the young' means that there are very few to take over the reins when the generation currently maintaining the diversity dies out.

With regard to the marketing of the Cooperative's products, one challenge has been the issue of quality and standards. Wholesale traders often find it easier to buy rice and other crops at the market, rather than directly from individual farmers. One explanation is that the traders fear that buying produce directly from the farmers might add to their costs because of uncertainties concerning the quality and regular supply. The assistance of organisations like LI-BIRD is therefore useful for creating relationships and promoting trust between farmers and traders.

LI-BIRD has also helped the farmers and processors meet the needed criteria. However, it has proved challenging to meet some of the standards, particularly with regard to uniformity. When marketing a variety of products based on agricultural biodiversity produced in small quantities by various smallholder farmers, it is difficult to deliver the uniformity and large volumes that the conventional market normally deals in. Selling on the international market would be even more challenging, as the requirements would be even stricter. Farmers' groups need alternative markets for their products. This is partly what LI-BIRD has tried to achieve, by starting small and focusing on the domestic market. The goal is for the market chain to be able to function well without project support.

As a possible long-term approach to promote market mechanisms that support on-farm conservation of agricultural biodiversity through the processing, distribution and sale of small-scale products, LI-BIRD is planning to invest in a private company. The intention is for most of the profits to go to small-scale farmers.

While the Pratigya Cooperative was not very active when LI-BIRD first came into contact with it, its members have now become quite confident and self-reliant. In fact, the Cooperative no longer needs active support from LI-BIRD – and that is an important indicator of the success achieved. Another indicator of the Cooperative's success is the improved livelihoods of its members. As we saw from the story of Maina Thapa, by linking conservation of agricultural biodiversity with development efforts it is possible both to create enthusiasm for diversity management and to improve livelihoods. It therefore provides a very useful example of how Farmers' Rights can be realised through benefit sharing.

Farmer breeders and participatory plant breeding

Bringing older varieties back to the market is not the only approach to maintaining diversity that has proven fruitful in this area of Nepal. When the older varieties contain both useful and less useful traits, the most important thing for farmers in order to benefit from the diversity can be to maintain the beneficial properties while reducing the impact of the less useful or even harmful traits. Enabling farmers to do their own breeding and actively engaging them in participatory plant breeding (PPB) can contribute to this. A PPB group in Sundaridanda village in Begnas that has received support from the In Situ project and collaborates with LI-BIRD regarding on-farm conservation of local crops, illustrates how such initiatives can contribute to the realisation of Farmers' Rights. Central to this group is a local farmer breeder, Surya Adhikari.

A farmer breeder's story

Surya Adhikari first became interested in breeding when he learned from the scientists involved in the In Situ project about the history of rice and how all cultivated rice varieties originated from wild rice. Through this project, LI-BIRD started working in the area in 1998. Up until then, the wild rice of the area had been used only as animal fodder (see Karki, 2004), but after being inspired by the project Mr Adhikari wanted to use wild rice to breed new rice varieties that could be well-suited to waterlogged low-lying fields. However, as he did not know how to cross these different types of rice, he asked a LI-BIRD plant breeder involved in the In Situ project for help. He was then given a three-day course on rice breeding – learning to cross and select segregating lines. Afterwards he was able to start his own breeding efforts.

In 2000/2001, Mr Adhikari did his first crossings and since then he has continued his work. Altogether he has made five crosses so far. He mostly crosses local varieties with wild rice and has also tried to use modern varieties. Some of his lines have proved especially promising and have been bred for multiple generations, whereas another cross was dropped due to high disease susceptibility. For the testing and selection of the various lines, Mr Adhikari receives assistance from the PPB group. This group, together with LI-BIRD, also helps him access land suitable for rice cultivation, as he does not own such land himself. When his eyesight deteriorated and his hands started to tremble some years ago, he helped his wife, Saraswati Adhikari, learn the necessary techniques and related knowledge. Since then she has been responsible for making the crossings and assists Mr Adhikari in his plant breeding work.

Mr Adhikari is very keen on developing new varieties and sees plant breeding as a good way to contribute to the national heritage. He also appreciates the fact that his plant breeding efforts have brought him in contact with many different people. Through participation in national and international forums he has had the opportunity to exchange information and knowledge with other plant breeders and farmers as well as other stakeholders. One of the most important benefits he has experienced from his plant breeding so far is therefore the ability to share something others value and the social recognition he receives because of his work. He is now also teaching other farmers his skills. Though Mr Adhikari has not yet benefited financially from his plant breeding efforts, he is quite happy and satisfied with his work and contribution to plant breeding.

Mr Adhikari thinks that the time and effort he has put into developing new rice varieties should grant him the right to be recognised if his varieties are registered. For him this means that the name and listing of the variety should reflect his involvement and the involvement of anybody else that might have taken part. He cares more about such recognition than monetary benefits because he views his varieties as contributions to Nepal's development. He also thinks similar principles should be taken as the point of departure for the registration of varieties bred by plant breeders who are employed and paid by the state. As their work is done for the benefit of the country as a whole, they should not be granted any particular personal rights to the varieties they have bred. In his opinion, recognition is important in such cases as well, but there should be no exclusive rights.

Although monetary benefits are not what he cares most about and he generally feels that the varieties he breeds are part of the common heritage of the Nepalese population, Mr Adhikari does think that he should be entitled to a share of the profits if someone uses a variety he has developed for commercial purposes, for example by selling seed. However, he wants other farmers to be able to use the varieties freely.

Activities of the PPB group

Mr Adhikari serves as chair of the PPB group. This group, established in 2003, had 25 members as of late 2012. Coordination and management is taken care of by an Executive Committee consisting of nine elected members. When the group was formed as part of the In Situ project, the goal was to involve local farmers in PPB and enable them to take the lead and be responsible for decision-making. Growing the early lines of a cross is not something all farmers are interested in doing, as yields tend to be uncertain and low. In addition to an interest in plant breeding, good observational skills are needed. In order to establish the PPB group, interested and research-minded farmers were identified. These farmers then received training in the principles and techniques of plant breeding, particularly concerning selection and advancement of segregating lines in subsequent generations, and maintenance of breeding lines and seeds.

Initially, the PPB group received help with conducting the crossings. Once the crosses are done, the group plants the seeds in farmers' fields, then evaluates and selects promising or preferred lines. The first step in this process is usually a group meeting to select members with suitable fields for planting the seeds of breeding lines. Seeds are given to these farmers, and other members provide labour support during planting. Each field planted with breeding lines is visited by the entire group twice each season; first during the vegetative stage and then during the ripening stage. Other farmers from the community are also invited, and together with breeders from LI-BIRD the group selects plants with farmer-preferred traits in each generation. When the breeding lines have reached the third or fourth generation, the group normally decides which line they want to proceed with. Then the evaluation and monitoring continues. By the sixth or seventh generation, the preferred lines usually become stable, in terms of plant types, yield and other traits. At this point, the group will treat the line as a new variety and maintain it separately. Seeds of these stable advanced lines are distributed to interested farmers in the village as well as in the neighbouring villages. The advanced lines preferred by a large number of farmers are then selected for registration as new varieties.

One of the varieties the PPB group chose to focus on was *mansara*. This is a local rice variety that performs well in dry environments, but its straw yield, disease resistance and taste are relatively poor. The PPB group decided that they wanted to improve this variety, aiming to retain its positive features while also improving its yield and eating qualities (Gyawali *et al.*, 2006). The approach they chose was to cross it with a modern variety called *khumal 4*, a high-quality rice with good eating qualities as well as relatively long straws and good straw yields. After starting out with more than a hundred different lines and after many years of testing and evaluation, the PPB group is now down to five lines. Two of these – *mansara 4* and *mansara 5* – are actually among their most promising lines altogether. As the result is drought-resistant varieties with high straw yield and desirable eating qualities, the group is very pleased. Compared to the old *mansara*, these new varieties have a 30 per cent higher yield. The group is planning to register these two varieties, *mansara 4* and *mansara 5*.

The PPB group has also been working with lines from other crossings. Here the most promising are two lines from a crossing between *biramphool* and *himali*. The former is an old local variety with generally low yields; the group wanted to make it tolerant to lodging and increase its yield while retaining its good taste qualities (Gyawali *et al.*, 2006). The continued selection done by the PPB group has resulted in two new varieties: *biramphool 3* and *biramphool 6*. These have better grain and straw yield, fewer lodging problems, higher resistance to disease, as well as finer grains and aroma than the original *biramphool;* moreover, they mature at the desired time. As these new varieties have high-quality traits like fine grain and aroma, farmers get a higher price for the grain, in addition to the improved income due to the higher yields. The group is planning to register also these varieties.

The PPB group has also contributed more directly to the conservation of endangered local rice varieties. When it was discovered that four varieties of rice (*yekle, mansara, anga* and *kathegurdi)* were being cultivated by very few farmers and the In Situ project had classified these varieties as endangered, the group distributed seed to its members and other farmers. This on-farm conservation effort has proven successful, and more farmers are now cultivating these varieties.

Because of their own success, the PPB group is now recommending farmers in other areas and villages to work together in groups, to take part in PPB efforts and to focus on conservation and use of local agricultural biodiversity. They have helped establish similar groups in several areas and have provided them with advice. The group also plans to start seed production of the most promising varieties. The resulting income is to go to the group.

Government involvement

During the implementation of the In Situ project the PPB group worked closely with the Nepal Agricultural Research Council (NARC) and the District Agriculture Development Office (DADO) of Kaski District. Although the project has now been phased out, the group still maintains informal contact with these two institutions. Every year, the group invites plant breeders from NARC and DADO to participate in the joint monitoring of breeding trials and to provide technical advice and feedback. NARC has also provided help with regard to monitoring of pests and diseases.

Similarly, DADO is willing to help test and spread the PPB varieties in other parts of the district. They have already begun doing so for the new varieties of *mansara* and *biramphool*. The new *mansara* varieties, in particular, are believed to have great potential in other dry areas.

There is, however, a need to formalise and institutionalise the PPB group's partnership with NARC and DADO to further strengthen its work. The group members themselves think it should be government policy to provide technical and financial support to groups like theirs since the work of such groups helps maintain Nepal's agricultural biodiversity and enhances the livelihoods of smallholder farmers.

Challenges

One of the challenges related to the breeding has been costs. Advancing and maintaining breeding lines can be an expensive process, not least because the members have to set aside land for evaluation and testing, and if there are problems the farmers will also experience a yield loss. The costs in terms of time spent are not regarded as an issue to the same extent as the farmers realise they are doing the work for their own benefit.

According to the Chief District Agricultural Development Officer of Kaski District, Beni Bahadur Basnet, one of the challenges with regard to agricultural development and farmer involvement in Nepal is that there are no organisations that really represent farmers. Instead, all the political parties have their own farmer unions, which are used to mobilise the farmers for the parties' own purposes, rather than channelling the views and demands of farmers. The PPB group would like to create regional groups for farmer breeders and PPB groups and then organise these in a national organisation. Organising farmer groups in this manner is being supported by LI-BIRD.

Benefits of PPB and group work

Members of the PPB group have found that both the results of their breeding efforts and the group participation itself constitute important benefits. In addition to developing varieties that they will benefit from growing, members appreciate the collaboration and how they learn from each other. Part of what holds the group together is the recognition that they can achieve more together than alone and the sharing of both the burdens and the benefits.

Testing requires access to various types of land, and here it is better to work together as a group. Spreading the testing fields in this way also means that a larger number of neighbours will be exposed to, learn about and be able to participate in the activities of the PPB group. The group also appreciates the opportunity to evaluate different plant varieties in-depth.

Group members view the increase in yields as something that benefits not only them but also a much wider group of people. Members are proud of themselves, their achievements and the knowledge they have gained, and they now have a new perception of themselves as farmers. As a result of their work, they have also received recognition from various stakeholders, and this recognition is viewed as a benefit in itself. The members are also pleased to be less dependent on external varieties that are often less adapted to their fields.

In addition, the PPB group is also satisfied with the impact the In Situ project in general has had on conservation in the area, as it has both created awareness around conservation and halted the loss of diversity. They believe that many local varieties would have been lost if it had not been for this initiative.

The story of the PPB group and Mr Adhikari is therefore a good example of how Farmers' Rights can be realised through benefit sharing when the farmers themselves take part in breeding and conservation concerns are integrated in their breeding work.

Common concerns

According to Mr Basnet, the key to succeeding with conservation efforts is to make conservation profitable for farmers. In his experience, many farmers ask about the purpose of conservation, so he feels that an important part of conservation projects should be to raise awareness and focus on making conservation profitable for those involved. As he sees it, a good approach is to identify the most valuable local varieties and their traits and then improve them in a way that will make it profitable to cultivate them.

In this connection, Mr Basnet also underlined the importance that his office, DADO, places on coordination and cooperation with LI-BIRD and other NGOs to avoid overlapping of efforts. He tries to ensure that DADO-initiated projects support and complement LI-BIRD's projects. Linking up with NGOs is also important because of DADO's budget constraints. In his view, one of the challenges in Kaski district with regard to development projects is the high number of NGOs operating in the area and the lack of coordination. More than 30 NGOs have been working on projects related to agriculture in the district. To get an overview of the situation they were all invited to a meeting by the DADO in 2008, not least to find out who was involved in what and where.

This process was quite revealing. It turned out that in one part of the district there were as many as 13 NGOs involved and that they knew little about each other's activities, whereas in other parts of the district there were no NGO projects at all. The general tendency was for the NGO projects to be clustered around Pokhara and scarcer in remote areas. As a result of the lack of coordination there was also some overlapping of efforts. To address this problem, Mr Basnet asked the NGOs to start sending him their project descriptions so his office could help avoid any overlapping and too much clustering of projects.

Farmers in the area were also given the opportunity to share their views at a meeting, on what they wanted from NGOs and the government. These wishes were then presented to the NGOs working in the district. Some NGOs, among them LI-BIRD, then decided to focus on the aspects highlighted by the farmers.

As is the case in many other regions and countries, the younger generations in Kaski district in Nepal are also showing less interest in farming. The Pratigya Cooperative struggles in recruiting younger members and they are not alone in facing this problem. Many young people from the rural areas in the district are leaving agriculture behind to go to the cities or abroad. This will probably continue as long as neither they nor their parents see any future for them at home.

However, although the young might be leaving farming, the behaviour of consumers is changing. Greater appreciation of local varieties can be seen, and local crops and varieties are now increasingly being integrated into the local economy. This gives hope for the future, as possibilities to improve farm livelihoods while at the same time maintaining the local diversity heritage might tempt some of the younger generation to take up agriculture.

However, many challenges still remain for projects like those described here. For one thing, it will be important to scale up successful approaches so more farmers can become involved in and benefit from the maintenance of agricultural biodiversity, making it possible to realise Farmers' Rights on a larger scale.

The way forward

In order to have an impact on more farmers and contribute to realising Farmers' Rights on a larger scale, the central principles must be brought to other areas as well. LI-BIRD is therefore working on scaling up its activities. The potential for impact is great; and if successful, such efforts will offer important lessons for all those working on these issues. In their evaluation of their own projects in Kaski district, one of LI-BIRD's conclusions was that there was a need to legitimise farmer organisations engaged in on-farm conservation of agricultural biodiversity in the area. They also felt it was important to take the existing administrative structures as the point of departure to a greater extent. While LI-BIRD previously preferred to work with existing groups and found such groups to work with, as in the case of the JSSA and the Pratigya Cooperative, in their new organisational model for farmers' groups they utilise Nepal's administrative structure and have facilitated the creation of Agricultural Development and Conservation Committees (ADCCs) in Begnas.

This new model has been pioneered in Rupakot VDC. Rupakot VDC consists of 14 villages and comprises about 800 households. All the farmers in the area are free to join the ADCC, and on each administrative level the ADCC collaborates with the corresponding local administration and elected officials. The nine wards of Rupakot VDC all elect nine people to their own ward-level committee, and then these committees elect the 12 farmer representatives that make up the VDC-level committee. The ADCC tries to influence the VDC to use some of its funds on the agricultural sector and coordinates the activities related to conservation and management of agricultural biodiversity at the local level.

This new approach has proven fruitful thus far, so LI-BIRD has begun using it in its other project communities in Nepal. In addition, an *ad hoc* umbrella organisation consisting of the ADCCs established so far has been formed to enable farmer representation at the national level. Mr Surya Adhikari from Begnas has been elected to chair this organisation, and the Ministry of Agricultural Development has appointed him as a farmer representative to the National Agricultural Biodiversity Conservation Committee. This can be seen as a step in the right direction with regard to the right of farmers to participate in decision-making and therefore as another achievement that contributes to the realisation of Farmers' Rights. Capacity building for farmers and investing in farmers' organisations are important and should be given high priority in national and international policies and programmes aimed at on-farm conservation and sustainable management of agricultural biodiversity.

With regard to scaling up activities like marketing support, questions emerge as to balancing supply and demand and the competition that might arise among the various villages involved in the production and marketing of products based on local and traditional crops. LI-BIRD and the other involved partners are struggling to find solutions, as they try to bring the models that have worked so well in this part of Nepal to other areas.

Concluding remarks

This success story from Nepal shows the positive impact that projects focusing on conservation of agricultural biodiversity and benefit sharing can have and how Farmers' Rights can be realised at the local level through such an approach. The success of the initiatives in Kaski district has even influenced the national programme of the Nepal Agricultural Research Council: it now focuses more on participatory plant breeding and the use of local and traditional genetic materials in breeding programmes, than before.

We have seen how a national NGO with international partners has played a central role for the successful initiation and follow-up of these initiatives. This story from Nepal therefore showcases the important role NGOs and the civil society in general have played so far for the realisation of Farmers' Rights, particularly with regard to benefit sharing. However, it also points up the need for good coordination to avoid too much overlapping and clustering of activities in areas where a large number of NGOs operate.

Perhaps the most important lesson this success story has to offer concerns the need to ensure that the maintenance of agricultural biodiversity is *profitable for farmers:* only then will the majority of farmers actually be able to take part in such activities. Many of the farmers still engaged in maintaining local varieties are poor. Making sure that they benefit from their important contributions is very likely one of the most efficient ways of protecting these varieties from dying out, as environmental and financial pressures continue to mount. The realisation of Farmers' Rights through benefit sharing therefore has the potential to be one of the most important tools for securing the world's agricultural biodiversity.

Notes

1　Important in this context was the Nepal Country Component of the global project 'Strengthening the Scientific Basis of In situ Conservation of Agricultural Biodiversity On-Farm' coordinated by the International Plant Genetic Resources Institute (IPGRI, now Bioversity International). In Nepal, the NGO LI-BIRD and Nepal Agricultural Research Council were central.

2　This chapter is based on information from a series of four group interviews (with representatives of the ADCC in Rupakot, the JSSA, the Pratigya Cooperative and the PPB group, including Surya and Saraswati Adhikari) and five interviews with individuals (the Secretary of Rupakot VDC, Maina Thapa, the DADO of Kaski district, the owner of Namuna Gaunle Shop and the owner of Janaki Agro-products) conducted during a fieldtrip by the authors in the Pokhara area, February 2009. The information was updated in 2012. The authors would like to thank the LI-BIRD staff, particularly Bikash Paudel, Sudha Khadka and Shreeram Subedi, for all their assistance during the visit, as well as all the interviewees.

3   Nepal is divided into administrative districts, which in turn are divided into towns and Village Development Committees (VDCs). Each VDC is divided into wards, which may be composed of one or more villages, depending on their size. Begnas VDC is located in Kaski district; the district capital is Pokhara.

4   The two other organisations are Kishan Dekhi Kishan Samma, or KIDEKI ('from farmers to farmers') and Rupatal Punarsthpana Tatha Matshya Sahakari (the Rupa Lake Rehabilitation and Fishery Cooperative).

5   When the In Situ project initiated this community documentation, it was the first such initiative in Nepal. Project experiences have since been utilised by the Ministry of Forest and Soil Conservation in its work on developing national guidelines for such community registries.

10 Community seed fairs in Zimbabwe

Robert Chakanda, Andrew Mushita and Tone Winge

In January 2009, several rounds of droughts had crippled Zimbabwe's already struggling food production and the country was desperately trying to cope with the worst cholera crisis in Africa for 15 years – but the farmers of Uzumba Maramba Pfungwe (UMP) district still managed to organise their annual seed fair. This fair was part of a series of annual seed fairs organised by farmers in Zimbabwe with the help of the Community Technology Development Trust (CTDT), a Zimbabwe-based NGO, that have promoted Farmers' Rights by providing farmers with the opportunity to display seed from their own varieties and access seed from other farmers and regions. In this way the seed fairs facilitate the right of farmers to exchange and sell farm-saved seed and provide an incentive structure for the conservation and sustainable use of plant genetic resources for food and agriculture.

According to Dorothy Chiota, spokesperson for the UMP women's group for gene-bank management and seed fairs, the farmers all look forward to these seed fairs. On average, the annual seed fairs attract about 2,000 farmers. Most of them come from the surrounding villages, but many also travel from villages farther afield. Seed fairs have been organised in UMP since 1997. They were initiated by the CTDT, but the local farmers are now the main organisers, with the CTDT functioning in more of a supportive role.

This chapter tells the story of how these successful, farmer-driven seed fairs are organised and of the positive impact they have had on the maintenance of agricultural biodiversity and Farmers' Rights in Zimbabwe – even in drought-prone areas like UMP, and during difficult periods in the country.

Legislation relevant to Farmers' Rights in Zimbabwe

There are several laws in Zimbabwe that affect farmers and their maintenance of local genetic resources. The Seed Act of 1965 regulates the production of commercial seed by seed companies but not the production and sale of traditional varieties. In practice, farmers are allowed to produce and sell seed of registered varieties, while landraces and traditional varieties are to be sold only by seed companies. These somewhat unusual rules are not specified in the Seed Act but constitute part of how it is implemented. Because the commercial seed companies tend not to focus on local and traditional varieties, such seed is not easy to obtain from the formal markets, and it is usually difficult for communities in one area to obtain local varieties from communities in other areas. The improved access that the seed fairs provide is therefore an important benefit for Zimbabwean farmers.

The CTDT considers the rules governing seed production and distribution in the country to be detrimental to the rights of small-scale farmers. These farmers usually prefer to sell their seeds independently, without having to go through the established seed companies; also, most farmers want to sell their seeds openly as seed and not as grain. According to the CTDT, current legislation in Zimbabwe favours large seed companies, while it restricts the local farmers and does not address their needs. In the organisation's opinion these laws still bear the mark of colonial rule, and changes are therefore seen as necessary.

In addition to the Seed Act, the Plant Breeders Rights Act, last revised in 2001, also affects the implementation of Farmers' Rights in Zimbabwe. This act limits the rights of farmers to re-use and sell seeds of protected varieties for the purpose of reproduction and multiplication. Farmers who cultivate less than 10 hectares of land are allowed to re-use farm-saved seed of protected varieties from their own holdings on their land but not to sell or exchange that seed. However, farmers who get 80 per cent or more of their annual income from farming on communal or resettlement land may multiply the seeds of protected varieties and exchange the seed with other farmers in this category.

Despite these exceptions, the Plant Breeders Rights Act does restrict the rights of farmers to sell, re-use and exchange seeds of protected varieties and the CTDT feels that it limits farmers' access to plant genetic resources. The sharing and exchange of seed has traditionally been important in Zimbabwe. Seed that has passed from farmer to farmer through generations is regarded as common property. Further, according to the CTDT the barriers to traditional practices created by the current legislation have contributed to the growing dependence of many farmers on hybrid varieties that provide higher yields only if expensive inputs are applied and if rainfall has been sufficient. As a result, local varieties are gradually disappearing from the farms. The CTDT wants to counter this trend by reforming the current system and ensuring that the country's biodiversity can be maintained and the local farmers can benefit from their contribution to the world's agricultural heritage by having access to a wide range of seed.

The CTDT's involvement in Farmers' Rights

Since the CTDT was founded in 1994 it has worked on agricultural biodiversity, food security and rural development in western and southern Africa. The concept of Farmers' Rights has become important for the organisation and it has tried to influence the government of Zimbabwe to enact laws that support farmers' access to and control over crop genetic resources. The organisation's engagement in policy work related to genetic resources dates back to 1999. At that time, a patent application by a professor at the University of Lausanne in Switzerland related to the medical plant *Swartzia Madagascariensis*, and in the CTDT's opinion, based on traditional knowledge from Zimbabwe,[1] made clear the country's need for a legal framework regulating the access to biological diversity.

In response to the *Swartzia Madagascariensis* case the CTDT started to organise consultative processes with a range of stakeholders and to explore the need for *sui generis* access legislation that could protect the interests of local communities. These processes culminated in the development of a draft regulation on access to biological diversity that covered access to wild and agricultural biodiversity, as well as the protection of indigenous knowledge systems and Farmers' Rights. This draft was presented to Zimbabwe's Ministry of Environment and Tourism.

The draft act was not adopted in its entirety, but sections of the proposal were incorporated as regulations under the Environmental Management Act. However, Farmers' Rights were not among the issues included in the adopted act. As a result, CTDT has continued to lobby for the enactment of legislation that deals with Farmers' Rights and fulfils the obligations of Zimbabwe as a party to the Plant Treaty. As part of this work, the CTDT in 2007 initiated awareness-raising workshops and dialogues with farmers and policy-makers, including the Ministry of Land and Agriculture.

In addition to its policy and legislative activities, the CTDT also wanted to work together with farmers on the practical realisation of Farmers' Rights and therefore initiated community seed fairs. Seed fairs can promote Farmers' Rights in various ways: they support traditional seed exchange practices and provide improved access to a wide range of seed. As Zimbabwean farmers were in need of better access to genetic resources, the seed fairs showed promise in terms of promoting benefit sharing. In addition, the focus on diversity can promote the conservation and sustainable use of crop genetic diversity.

The CTDT has community development programmes in nine districts of Zimbabwe: Mutoko, Murehwa, Goromonzi, Nyanga, Chegutu, Chiredzi, UMP, Tsholotsho and Mudzi. In three of these districts – UMP, Chiredzi and Tsholotsho – annual seed fairs are arranged in collaboration with the national Genetic Resources and Biotechnology Institute (GRBI). In both UMP and Chiredzi one seed fair is organised for the entire district, whereas three separate fairs are organised in Tsholotsho because of the distances involved. After the district-level fairs, an annual national fair is held.

Objectives of the seed fairs

A seed fair will traditionally offer a venue for local communities to display the crops they grow. The seed fairs initiated by the CTDT in Zimbabwe allow farmers to display their seeds and products, and all stakeholders may buy, sell and exchange seeds. In addition, the fairs provide participants and visitors with opportunities for interacting with other farmers as well as with a broader group of stakeholders. Farmers from the districts where seed fairs are organised can come together at least once a year to share ideas and experiences. In addition, the seed fairs provide the CTDT and extension staff from the Ministry of Agriculture with an arena to reach farmers and disseminate information.

Figure 10.1 Crop diversity at a seed fair exhibition
Source: Robert Chakanda

One of the main objectives of the seed fairs initiated by the CTDT is to promote access to local crop diversity and to enhance gene flow through seed exchange, knowledge sharing and technology transfer. The improved access to seed provides options for the recuperation, restoration and enhancement of local crop genetic resources. By enabling farmers to access cost-effective seeds of adaptable crop varieties, the seed fairs also contribute to increasing food productivity and food security.

A further important objective is for these seed fairs to serve as an arena for multi-stakeholder interaction, with knowledge sharing among scientists, policy-makers, farmers, extension agents and development practitioners. In addition, the fairs are intended to contribute to enhanced technological innovations by farmers and to provide incentives for competition, higher productivity, conservation and sustainable use of local genetic resources while minimising crop diversity losses. And finally, the seed fairs are meant to promote farmers' control over plant genetic resources, to empower farmers and raise their awareness of their rights in relation to seed.

Conserving genetic resources through seed fairs is not only a question of maintaining the local crop varieties for future generations: diversity is also a matter of shorter-term importance, vital to the daily food needs of smallholder farmers. Access to this pool of plant genetic resources provides a buffer for the poor to withstand the ravages of drought, climate change, instability and war. This is why it is such an important benefit. Often it is indigenous seed systems, not the global seed market, that provide food security in marginal environments where growing a range of nutritious crops can provide farmers with insurance against crop failures in periods of natural disasters like drought.

Seed fairs can also be crucially important to the conservation and sustainable use of plant genetic resources because they facilitate the registration and monitoring of local crop diversity. Information on the existence of crop specific diversity, threats and levels of decline can provide essential guidance for policy and management decisions at local and national levels.

In Zimbabwe as in many other countries, genetically uniform, modern crop varieties are increasingly being adopted, at the expense of more diverse, locally adapted varieties. Community seed fairs can help to diversify the range of crop varieties available to farmers and can play an important role in the identification of economically viable crops. They can also help in identifying varieties with specific dietary values. Such varieties might still be in use by farmers despite lack of support from the formal sector, and the seed fairs can help to make them better known. In addition, seed fairs can bring out important knowledge and information necessary for the sustainable maintenance of different varieties. Keeping this traditional knowledge alive requires the conservation of diversity at the farm level.

When these objectives are realised and the seed fairs improve farmers' access to a diverse range of varieties and the associated traditional knowledge they can promote benefit sharing. More specifically, seed fairs enable and encourage farmers to grow a diversity of locally adapted crops, and in that way they can show what national incentive structures might look like.

Organising a seed fair

At the seed fair held in January 2009 in UMP district, Dorothy Chiota described how such fairs are organised and gave her views on the benefits they have brought to her community. Ms Chiota is a member of the UMP seed fair committee; a group of selected farmers that also manages the community seed bank. She explained that when the CTDT first visited her area in 1998 and the local farmers were asked to participate in community seed fairs, they were told that these fairs would give them the opportunity to show their seeds to the public and enable them to see and obtain new seeds from other farmers and regions. In the beginning the farmers were hesitant, but after the first seed fair had been held in 1999, farming communities in the region were eager to continue with such events.

After a few years, these seed fairs became annual after-harvest events. Participation increased with more government representatives and other NGOs in addition to the CTDT becoming involved. National and international seed companies display new varieties and sometimes also tools and technologies that have been developed for production and conservation, and the farmers display their seeds. Even though this is a seed fair, farmers from the surrounding villages also take advantage of the occasion to bring cattle and other farm animals.

In Ms Chiota's area, farmers' groups are central to various agricultural activities. These groups were taken as the point of departure when the seed fairs started and now form the basis for village participation. Each participating village has at least one farmers' group and some larger villages have more. Each group member contributes with his or her crop varieties, so together the group is able to present considerable diversity at the fair.

Ms Chiota said that in her village almost everybody attends the fairs:

> We are all farmers, and our children accompany us. The children often help their parents, for example by bringing the seeds, mats and other equipment to the site. They participate in the activities like everyone else, both answering questions and asking questions of their own. When we wish to sell or exchange seeds, the children are always very instrumental – why not? They are the future of our communities and heritage.

There are usually more women participating in the seed fairs than men, and Ms Chiota explained that women are the keepers and custodians of seed. The women also tend to arrive earlier at the venue to prepare the seed stands, while the men arrive later.

Village-level promotion groups have been established for organising the seed fairs. These groups consist of traditional leaders, elders and CTDT staff. The promotion group first invites the farmers to general meetings, usually asking representatives from all the surrounding villages. Prior to the fair in 2009, for example, the leaders highlighted the objectives and explained the advantages of the initiative. Most of the organisational details are outlined at these meetings – such as the village where the seed fair will be held and how the day is to be organised, including which farmers' group should occupy which seed stand and what sections of the display grounds.

Ms Chiota explained that quite often the organisers from the CTDT take the back seat in organising the event. This was important to the farmers, she said, because it increases the understanding that the fair is actually for them:

> We own the seeds that we display, we know what we want to cultivate, we know how to cultivate these seeds year after year, and we own the farms. We make the organizers know this, and we are proud of this sort of ownership.

She and her neighbour are members of the committee that ensures that the seed fairs are conducted according to the agreed plan for the men's and women's groups. They also make sure that there is no overcrowding at the stands. Her neighbour is very proud of this responsibility, and Ms Chiota says that it instils both men and women with a sense of leadership because they are conducting their own affairs. The farmers also work together with a CTDT project team, who can provide advice when necessary.

As part of the preparations, a technical team of farmers also goes out and invites farmers' groups from different villages in the region. They assign exhibition stands to the participants – deciding which groups should occupy which portion of the show field and which activities are to be undertaken. This group is also responsible for the entertainment events; local singers, school choruses, dancers and art groups ask the technical team for permission to perform at the show.

The day before each seed fair the venue is inspected. Farmers are allocated stands around the community seed bank with the size of these stands varying according to the size of the farmers' groups. Thus, all farmers get a chance to display their on-farm diversity. The opening ceremony features speeches by district authorities, such as the District Officer, the District Agricultural Secretary and the Farmers' Union Secretary, and village representatives welcome any government representatives. For some of the larger fairs the news media are invited, including radio and television reporters.

At the seed fairs the farmers display the local varieties of a range of different crops, including groundnut, bambara nut, cowpea, tomato, castor, pumpkin, maize, sorghum, sunflower, watermelon, finger millet, beans and pearl millet. Sometimes seed companies will also display improved varieties, like hybrid maize.

According to Ms Chiota, the seed fairs are festive occasions with a lot of singing and dancing – which attracts many people from the surrounding villages. 'I like dancing, like many other women, and the songs remind us of our traditional values', she explained. Some of the songs that are sung extol the attributes and characteristics of the local crop varieties, like colour of the grain, taste of the food products, storage quality, drought tolerance, medicinal properties, nutritional qualities, resistance to pests and diseases and yield potential. Such songs function as tools to transmit knowledge from one generation to the next and are composed to praise the collective heritage of crop diversity because this diversity offers livelihood opportunities to the population. Ms Chiota notes that the young people are becoming more interested in singing and dancing and the songs have special traditional messages for them.

An important part of the seed fair is the discussion session where the farmers' groups come together and can ask questions about varieties that look new to them. Everybody attending the fair is encouraged to join these discussions. Knowledgeable farmers will talk about their new varieties, about their new practices in seed management and about the pitfalls to be avoided. If the discussion group becomes too large, the organisers split it into two or three smaller groups. The CTDT staff and representatives from various government units use the discussions to talk about the local varieties and the advantages they offer.

After the discussions, the farmers who have displayed seeds at the fair are given certificates of participation. A group of judges, consisting of both men and women, has been chosen to do an assessment of the displays according to certain criteria: number of crops, number of crop varieties, seed health, seed quality in terms of average size, colour and shape, special and unusual varieties. The judges often inspect the displays during the discussion session. Awards are then given to the individuals or groups whose displays achieve the highest scores.

The involvement of the Genetic Resources and Biotechnology Institute

According to Mr Kudzai Kusena, Acting Curator for the GRBI (formerly the National Gene Bank), the issue of Farmers' Rights has not been fully incorporated

Figure 10.2 Discussion session at the UMP seed fair
   Source: Robert Chakanda

into Zimbabwean policy and legislation so far. In his opinion, the lack of such a national policy has resulted in commercial agriculture based on genetically uniform varieties being more lucrative than the cultivation of local crops and varieties and this is a cause for grave concern. However, efforts are underway to develop a national policy framework that incorporates Farmers' Rights.

Mr Kusena explains that traditional farmers in Zimbabwe practise their rights to seed in a very natural way through seed exchange, without being fully aware that this is what they are doing. He therefore thinks it is good for farmers to be directly engaged in initiatives like seed fairs that highlight the issue of access to seeds and how it can be improved.

The GRBI is organised under the Ministry of Agriculture, and the fact that the ministry has approved the institute's involvement in the seed fairs and other campaigns targeted at raising farmers' awareness indicates that the ministry is not opposed to such projects. Even more direct support from the government came in 2010, when Vice-President Joice Mujuru attended the UMP seed fair. She remarked that she was 'quite surprised and overjoyed to see a range of indigenous food crops that we used to eat when I was growing up', some of which she had not seen for years. The Vice-President was particularly impressed with how the farmers had managed to maintain so many varieties in a drought-prone area at a time when so much biodiversity was disappearing. According to Mr Kusena, 'the occurrence of persistent droughts in recent years and the aspect of food insecurity has been the turning point for getting the policy-makers on their toes to push for the protection of farmers' varieties countrywide.'

Mr Kusena sees the GRBI as more than a custodian of seeds and genetic resources: it is an important partner in projects targeting rural communities, and its staff represents the government in awareness campaigns with NGOs, like the CTDT and the Biotechnology Trust of Zimbabwe. Together these institutions highlight key issues and work with rural communities through events like the seed fairs. There is a long-standing relationship between CTDT and the GRBI in this regard, and they have developed joint proposals and compiled analytical reports for the government's consideration. As part of its close cooperation with the government, the CTDT has also been represented on the board of the National Plant Genetic Resources Committee since 1997.

The greatest danger facing farmers today, as Mr Kusena sees it, is the growth of a system that gives ownership over genetic resources to large commercial entities at the expense of small-scale farmers. Patent laws and trade restrictions have not been beneficial to rural farmers, he says, and it is critical to protect farmers against exploitation. Not only the restrictions on the rights to save, exchange and sell farm-saved seed but also the idea that their genetic resources are inferior to modern varieties is disadvantageous to Zimbabwe's farmers, he explains.

Together, the CTDT and the GRBI have tried to place the advantages of farmers' varieties and the rights of farmers on the agenda. The GRBI has also supported the CTDT in developing various community seed banks in an effort to improve farmers' access to seed. The farmers are very proud of these seed-storage facilities: they recognise that they have been maintaining their local varieties for a long time, that these varieties are very useful and that it is important to acknowledge their value.

For the GRBI, the ultimate goal in working with the CTDT is to reach a point where the relevant government departments and ministries understand the meaning of Farmers' Rights and the importance of Farmers' Rights realisation for the conservation of agrobiodiversity. They also want to raise awareness of Farmers' Rights and related issues at the national level. It can be difficult to carry out consultative meetings in rural communities. The barriers are many, ranging from bad roads and poor telecommunication networks to failing healthcare systems and an uncertain economic future in general. As Mr Kusena sees it, 'the major challenge for the whole process is to get access to the resources and human capacity needed to conduct awareness campaigns at the national level within the current political and economic situation that Zimbabwe is facing.'

The impact of the seed fairs

The annual seed fairs organised by farmers in Zimbabwe in collaboration with the CTDT have contributed to increased crop diversity by facilitating exchange and sale of locally grown seeds and by providing incentives for the continued main-tenance of local varieties. Initiatives like seed fairs help to build the local seed distribution systems in a way that *ad hoc* aid interventions fail to do. Through these seed fairs, farmers have learned about varieties and traditional knowledge previously unknown to them and commercial seed producers have been able to learn about farmers' needs, tastes and concerns. The fairs have also created market linkages that might promote processing and value addition. Moreover, they have provided arenas for discussing seed quality and price.

As the CTDT sees it, the main objectives of the seed fairs – to promote the exchange of seed from local varieties and assess the crop diversity maintained by local farmers – have been achieved in the districts where these fairs are organised. The seed fairs have also identified participants who can produce large quantities of seed of specific varieties for sale or exchange. Because these fairs are local arrangements, it is relatively easy for farmers to buy large quantities of seed and transport them home.

Mr Kusena thinks the seed fairs have been very successful in engaging farmers and making them aware of the disadvantages they will face if the locally adapted varieties are lost. The success of these fairs can be measured in practical terms; he notes that 'in Hwedza district the farmers, after having attended the seed fairs organized in the area and receiving the starting materials there, have become interested in cultivating local varieties of sorghum, pearl millets, cowpeas and taro'. He also underlined that in Murehwa district, where the food-security situation was very bad for several years, average levels of food security have risen after the seed fairs started.

The seed fairs have also enabled the GRBI staff to identify and collect crop varieties not previously represented in their collection, and varieties held in the gene bank have been re-introduced to the communities during the fairs. The GRBI has also received direct requests from farmers for specific varieties, like the local sorghum varieties *Mutode, Chihumani* and *Mukadziusaenda,* whose seeds are then multiplied and distributed.

Dorothy and the other participating farmers in UMP think that the seed fairs have brought many benefits to their communities. In addition to the improved access to seed, they appreciate the opportunities the fairs offer for identifying endangered varieties so that measures can be put in place to save them. The farmers also feel that the seed fairs have made them more aware of their rights in relation to seed exchange and use. Additionally, they appreciate the recognition given to those farmers who cultivate the greatest number of varieties and crops, noting how the seed fairs in that way encourage healthy competition. They feel that the seed fairs have succeeded in encouraging other farmers to join the on-farm seed production system. Another important benefit is the exchange of traditional knowledge as to cultivation and cooking practices.

All in all, the improved access to a large number of varieties, many of them local varieties well-adapted to the local conditions, constitutes an important benefit for the farmers who attend the fairs. Moreover, this is a benefit that can translate into other benefits, such as improved food security. Although there is still some way to go before Farmers' Rights are fully realised in Zimbabwe, the involved institutions are proud of what they have accomplished so far. The CTDT and the GRBI estimate that some 60 per cent of the small-scale, traditional farmers in the districts where the seed fairs are organised have benefited directly from them.

Concluding remarks

Especially in countries and regions where it is difficult for farmers to access crop diversity through the formal market, community seed fairs can play a vital role. This success story from Zimbabwe has shown how improved access can be achieved.

One of the most important lessons here is that such initiatives must be farmer-driven and based on local ownership if they are to be sustainable in the long run and able to succeed under difficult circumstances. The ability of the farmers' groups in question to organise their seed fairs even in the midst of droughts and disease has shown the robustness and stability of this model. The CTDT was instrumental in initiating the fairs and still provides important assistance, but once the organisational structure of the fairs had been established, the organisation stepped back and the local farmers took control. This has probably contributed to the success of these fairs and shows how civil society organisations may contribute to the realisation of Farmers' Rights.

The active role of the GRBI, especially with regard to the registration, monitoring and re-introduction of varieties, has also been central to the success in Zimbabwe and highlights how government institutions might support such initiatives. The legitimacy accorded to the proceedings by this GRBI involvement has probably also contributed to the participation of seed companies and other stakeholders; helping to promote stakeholder interaction and seed access. According to the CTDT, the collaboration between the various stakeholders has been central to the success achieved.

The opportunities for interaction among farmers and between farmers and other stakeholders offered by the seed fairs have made a substantial impact on farmers' ability to access seeds as well as contributing to information exchange. Not only has the improved access to local varieties been crucial to the maintenance of this diversity, it has also contributed to increased food security for smallholder farmers as these crops are more adapted to local conditions and do not require expensive inputs.

For other countries and institutions interested in initiating similar seed fairs, it can be worth taking note of the discussion sessions and the awards featured at the community seed fairs in Zimbabwe. Discussion sessions can be very efficient in promoting knowledge exchange, which is so fundamental to the successful cultivation of new varieties. The awards explicitly recognise individual farmers for their contributions to the conservation of agricultural biodiversity and function as additional incentives to such work.

After some initial hesitancy among the farmers, the seed fairs quickly proved their worth. They are now very popular among the farmers and attract many visitors each year. As spokesperson Dorothy Chiota explains, almost everybody from her community attends and there is a particularly strong involvement of women. This popularity can be seen as testimony to the success of these seed fairs, which have brought enthusiasm and attention to the issue of genetic diversity.

Note

1   The professor had received the genetic resources from the University in Harare but according to the Berne Declaration the filed patent was not in accordance with the transfer agreement. Access and benefit sharing terms were then re-negotiated with the proper authorities in Zimbabwe, but negative results from the clinical trials made the University drop both the project and the patent request (see the website of the Berne Declaration for more information www.evb.ch/en/p14269.html).

11 Farmer innovation in Tigray, Ethiopia

Fetien Abay, Åsmund Bjørnstad and Tone Winge

Once upon a time there was a plant called Sorghum. As this plant was growing, another plant named Barley appeared. The Barley plant was much younger than the Sorghum, but it hurried to grow up and mature. When the Sorghum saw this it asked the Barley to wait so they could mature together, but the Barley said that it could not wait, because it had hungry mouths to feed at home. Thus, the Barley hurried up and matured before the Sorghum.

This little tale from the Tigray region illustrates the important role played by barley (*Hordeum vulgare* L.) in this part of Ethiopia. The Tigray region, located in the northern highlands of the country, is a major barley-producing area known for its long history of crop cultivation as well as its diverse geographic, climatic and socio-cultural conditions. Barley is a major crop, particularly in eastern, southern and central Tigray.

In the following, we will see that not only are specifically adapted traditional varieties currently in use in Tigray but traditional knowledge in varietal selection is still practised by farmers. The rich traditions regarding the cultivation, preparation and storage of barley are kept alive and expanded through *in situ* conservation, active use and innovation. Varietal selection and innovation by farmers are crucial to create a balance between crop productivity, household food security and the maintenance of plant genetic diversity. This example from Ethiopia also illustrates the impact that one farmer's breeding efforts can have and how the benefits from innovations based on crop genetic resources and traditional knowledge can be shared. External actors, in this case a research institution, can also play an important role in such efforts (see Abay, 2007).

Since 2002, local knowledge and practices have been strengthened and added to through a collaborative project on barley breeding with both local farmers and researchers from Mekelle University in Tigray. This cooperation has also benefited farmers outside the project area as the farmer-developed barley varieties have been introduced in other villages. The link between benefit sharing and the maintenance of traditional knowledge that can be seen here is an example of how the various aspects of Farmers' Rights can be jointly realised.

Crop diversity in Tigray

The Tigray region is characterised by low levels and high variability of rainfall, by variable soil fertility and low levels of external inputs. One way for farmers to protect themselves against such environmental variability is to grow crops and varieties that differ in terms of sowing and maturity dates, duration of growing period, drought sensitivity and susceptibility to pests and disease (see Almekinders and Elings, 2001). In this way they can reduce the risk of losing the entire harvest. How barley diversity is used by farmers in Tigray has been described by Abay *et al.* (2008). Areas with high environmental diversity and variability that are regarded as marginal are therefore often centres of diversity for many crop species. This diversity is central to the livelihood of the farmers who live there; moreover, it might nurture traits of use for other areas. It is therefore crucial to maintain this diversity, together with the associated traditional knowledge.

Although the Tigray region has experienced profound demographic, economic and environmental problems in recent decades, the local farmers have still managed to retain and enhance crop varietal diversity. The improved varieties from formal agricultural research have not performed well under the local environmental conditions, so farmers usually focus on varieties that are more specifically adapted.[1] On the whole, Tigray can be divided into two environmental zones – the lowlands and the highlands – and the varieties that have traditionally been grown in these zones have been specifically adapted to their environments. Such specifically adapted varieties can contain valuable traits; for example, Grando *et al.* (2001) have demonstrated their value as regards drought-resistance. Although dry areas often are considered less relevant for crop production, the majority of such areas, including the Tigray region, actually host a unique crop diversity that has evolved in line with the environmental conditions, indigenous practices and innovations.

Traditional knowledge and local seed storage practices

Storage of the barley to be used as seed the next season is essential to variety maintenance and household food security. The farmers in Tigray employ various storage methods and apply a range of criteria in choosing the best and most secure places for seed storage. Clay pots and cattle leather containers called *aybet* and *lokota* are among the containers most used for seed storage in Tigray. Also widespread is the use of *shirfa*, traditional containers made from bamboo splits, whose internal walls are plastered and smeared either with mud or cow dung and the top covered with a flattened mud or clay plate, and *godo*, made entirely of mud or cow dung. One reason such containers are used is because they take up so little space, and the container size can vary with the produce and space available in the household.

These containers and the preferences regarding when to use them are all part of a culture of traditional knowledge in Tigray. Such knowledge is central to farmers' ability to maintain the diversity in their fields: it is important that it is being kept alive, used and shared.

The local farmers have various ways of ensuring seed quality. Stored seed that shows signs of infestation or damage is used for consumption rather than being kept for seed stock. Individual farmers or farmer groups sometimes also experiment with new practices, such as burning goat manure into a type of powder that is mixed with the seeds or mixing barley with teff, to enable longer storage. A technique developed by a woman from central Tigray, where the leaves of a traditional tree are ground and the resultant powder is mixed with barley seeds before sowing also illustrates the development of new practices. This tree-leaf technique has proven effective as protection against termites and has been disseminated both within and outside the woman's village.

As these examples show, the traditional knowledge related to barley cultivation in Tigray is still evolving. Avoiding stagnation of traditional knowledge is central both to keeping it alive and for ensuring that the practices are being adapted to changing conditions. This is essential in a world where the impacts of climate change are increasingly being felt. Documenting and facilitating the sharing of traditional knowledge related to crop genetic diversity as well as pairing it with new technology and science as university researchers have done in Tigray, can be one way of promoting resilient traditional knowledge and thereby realising Farmers' Rights.

Varietal selection in Bolenta

Research suggests that farmers in Tigray may sometimes get higher yields if they practise seed selection (see Aderajew and Berg, 2006), which again can improve food security, provided seed health is also taken care of. In Tigray, farmer communities are quite familiar with varietal selection, and the selection processes known locally as *wulad* (first generation), *salisin* (third generation) and *argit* (old generation) are well-known. There is also a high degree of seed exchange among farmers of the region, both within and between villages; and the seed selection process is one of the main drivers of diversity in the farmers' fields.

The village of Bolenta is located in the Endamekhoni district of Tigray at an elevation of 3,000 metres and surrounded by high mountains. The area itself is hilly and mountainous as well as cold and rather windy, especially from October to December during the period known locally as *ashayta*. Although Tigray is known for the shortness of its main wet season, normally lasting from late June/early July to mid-August, this part of the highlands enjoys rainfall over a somewhat longer season, from mid-June until September. This zone also has a shorter, second rainy season, which enables farmers to grow barley twice. Unlike the mid-altitude barley-growing sites and highland areas with no additional short wet season, farmers in the Bolenta area use early-, medium- and late-maturing barley varieties. In the villages of the Tigray highlands, such as Bolenta, barley is cultivated mainly for household consumption – in contrast to the mid-altitude areas, where it is also sold or exchanged.

One of the barley-growing farmers in Bolenta is Mr Kahsay Negash. At 92 years of age, he was still a very creative experimenter as of late 2012. His land consists of the 1 hectare he owns and another 1.25 hectares he rents. As the owner of three oxen, two cows, three donkeys and 20 sheep, he is regarded locally as a wealthy farmer. Over a period of ten years, using the skills gained through continued experimentation with selection practices during his whole career as a farmer, he developed

Figure 11.1 Kahsay Negash in front of a barley field
Source: Fetien Abay

two varieties of barley that combine the agronomic, morphological and eating
preferences of local farmers. Although it is time-consuming to select plants, mark
them and sow them in separate rows as well as separating their seed at harvest time
and storing it, Mr Negash felt it was necessary for him and his family to do this in
order to develop a variety that was well-adapted to the land they had enriched by
using manure from the livestock.

When asked why he started doing varietal selection, Mr Negash explained that
because the environmental conditions in his area are changing he has to plan ahead
and find varieties that can adapt to the changing circumstances. He also said that he
could not let his mind be idle. His experience with and close observation of pest
infestations, his successful experiments on land rehabilitation resulting in better-
quality soil and the cookery skills of his wife, Mihret, gave him the incentive to work
on crop varieties. We may say that his motivation came partly from necessity and
partly from curiosity.

Mr Negash is eager to talk to scientists, other farmers and family members about
new ideas. His family takes part in the breeding, and Mihret Negash and his son,
Mebrahtu Negash, have been particularly central. Mebrahtu has taken part in
marking and selecting the plants, while Mihret plays a significant role in processing
the barley variety *Himblil* for making good-quality *injera,* a local bread. Her skill in
this regard was one of the factors that motivated Mr Negash to start varietal

selection for *Himblil*. Mihret Negash also examines the plants before the harvest and selects seeds after harvesting and threshing. In addition she is responsible for cleaning, drying and storing the seeds, tasks that locally are seen as the responsibility of women. The important role played by women in the maintenance of specifically adapted varieties is recognised in the village – as is clear from the local proverb, *sebeyiti zeybilu – zeri zeybiln – Hiwot zebilun*, which can be roughly translated as 'no wife – no seed – no life'.

In Bolenta, relatively strong social networks regulate and facilitate the exchange of information and seed. However, despite general familiarity with the selection processes, not all local farmers are fully steeped in the detailed knowledge behind the practice of seed selection and achieving good yields. Some farmers therefore think Mr Negash gets such good harvests because he frequently uses manure on his land (although in fact most farmers in the area do this) and has built stone terraces: in other words, they do not recognise it as the result of his work in plant breeding and selection. The physical structures of terracing have indeed reduced the problem of flooding for Mr Negash's family, and the organic matter from the manure has allowed more water to infiltrate. And as a result, Mr Negash's farmland and garden have benefited from soil moisture over a longer period. These factors have indeed contributed to his good yields – but at least equally important has been his advanced selection work.

Demhay

One of the barley varieties Mr Negash has worked on is a six-rowed variety of 'naked' barley called *Demhay*. Naked barley has received its name from its hulls, which due to a single recessive gene are so loosely attached that they fall off during harvesting. In Ethiopia, the tradition of cultivating naked barley is as old as that of cultivating covered barley, and it is one of the countries where naked barley is grown for human consumption. However, the formal agricultural research sector has so far devoted little time and attention to naked barley, even though it tends to have a higher protein content (see e.g. Oscarsson *et al.*, 1996) than hulled barley. According to the local farmwomen of Tigray, naked barley is more suited to food-preparation than covered barley as it does not have to be de-hulled.

Despite its advantages, the cultivation of naked barley is decreasing in Ethiopia because most of these varieties have lower yields than varieties of covered barley. Because relatively little is known about the source of drought-resistance in naked barley, it is difficult to obtain cultivars that yield well under drought-prone conditions. The search for local knowledge about drought-resistant varieties and how to cultivate them is therefore important for Ethiopian programmes that aim at conserving and breeding cultivars of naked barley.

Demhay is appreciated by the local farmers for the taste it brings when used for beverages, bread and the local snack known as *kolo*. *Kolo* is made of roasted grain, and the local farmers describe *Demhay* as the preferred variety for making *kolo* because of its good popping qualities. It is also preferred for making the local beer called *siwa* due to its good fermentation qualities and flavour. This last preference is reflected in traditional songs and sayings, such as: '*Siwakhen le siwa Demhay eyu − lihameme lehiyu*', which means 'your local beer is made from *Demhay* − it is so good it will cure the sick'. People in the area sing this during festivals and wedding ceremonies to point up the good qualities of *Demhay* beer. Such songs and sayings clearly illustrate the importance of barley in the everyday lives of people in Tigray and show that the knowledge about the various varieties and their properties is still an integral part of the local culture.

Demhay can also be used to make *tihlo*, a local dish consisting of doughy balls made from barley that are dipped in a spicy meat sauce mixed with yoghurt. It is made by kneading *tihni*, the fine flour of barley, into dough. Small pieces are taken by hand from the dough, rolled into small balls and served with a spicy meat sauce. Both the eating style and the preparation method of *tihlo* are different from that associated with *injera*, the bread normally made from teff. This illustrates the rich culture that exists around barley and how different varieties are cultivated for different purposes.

Mr Negash has been working on the selection of naked barley for a long time. As part of his interest in cultivating late-maturing varieties, he decided to select plants from *Demhay*. However, he has not given the resulting variety a different name because he thinks it can be easily identified by the black spot on its seed. According to Mr Negash, one of the distinguishing factors of his variety compared to the old *Demhay* is its improved drought-resistance. The old *Demhay* needs relatively high rainfall to produce high yields. Mr Negash's variety also differs from the three known varieties of *Demhay* − yellow, green and black − in that it prefers better soils, is of medium height and has grains shaped like those of wheat.

Mr Negash has supplied seed of his *Demhay* variety to interested farmers and given them advice on sowing dates and plot management, but because it needs better soils and more nutrients and has a longer growing period than the *Demhay* varieties normally grown in the area, Mr Negash's variety has not proven as well-suited to the land of other farmers as to his own land. However, another variety he has developed has been widely adopted by other farmers. This variety is called *Himblil*.

Himblil

Previously, Mr Negash's family sold barley to buy teff when they wanted *injera* for religious and cultural ceremonies, and they had to sell more than three units of barley to buy one unit of teff. However, the high teff prices and his wife's food-preparation skills encouraged Mr Negash to improve *Himblil*, a specifically adapted six-row barley variety, and his wife now prepares *injera* out of *Himblil* barley that is comparable in quality to the *injera* made from teff.

After having experimented with *Himblil* for a few years, Mr Negash noticed a few plants that differed from the rest in terms of spike characteristics and maturity. He decided to do selections based on these off-types and asked his wife and son to mark and harvest them separately. Central to this work was the soil fertility improvements Mr Negash had achieved and which enabled him to select single plants that stood out. His wife hung the plants over the kitchen hearth for drying, and Mr Negash sowed the seed the following year in a separate plot. He then continued to evaluate and multiply his seed stock: the result was a new *Himblil* variety.

This *Himblil* variety though based on the local *Himblil*, differs from it in many respects. The most important differences are related to straw yield, grain yield, plant height, number of seeds/spikes and utility-related traits, like suitability for *geat* (porridge), *kolo* and *kitta* (flat bread). In addition, *injera* made from Mr Negash's family's *Himblil* variety is comparable to the *injera* made from red-seeded teff in terms of texture and taste.[2] His family's variety is tall, has stiff straws, long awns and long spikes, and is thus unique. It matures earlier than *Demhay* but slightly later than the local *Himblil* variety. It also prefers soil of medium fertility but can adapt well to different soil-fertility levels and rainfall regimes.

Mr Negash has distributed this *Himblil* variety among the other farmers in Bolenta. Those who prefer his *Himblil* variety to local ones say that the decisive traits are its yield, stature, seed size and suitability for making *injera*. Although straw palatability is not the decisive factor with regard to choosing this variety over the original *Himblil*, this trait has brought substantial improvements as well. Farmers also appreciate his variety because of its early vegetative and grain-filling stage, compared to the longer vegetative stage of introduced varieties and late-maturing local varieties.

The degree to which Mr Negash's *Himblil* is recognised as 'his' by the other farmers in his village became clear when Mr Negash gave some seeds of it to a relative in a nearby village because he wanted to see how it would perform outside Bolenta. When his fellow villagers observed someone growing the variety in another village, they thought it might have been stolen from Mr Negash's field and told him about it. Explaining that he himself had given the seed away, Mr Negash was encouraged by the recognition his community gave to the variety he and his family had developed.

Developing his own version of *Himblil* as well as giving away seed, has indeed provided Mr Negash with recognition from his community. Mr Negash has given away seed from this variety to those who wanted it, and this particular variety of *Himblil* has spread and become quite popular. When many of his fellow villagers lost all of their seed in 2000 due to rodents, he provided seed to more than 40 farmers. In his opinion, his varieties have proved to be well-adapted to the local environment. Mr Negash's generosity in providing seed from the varieties he has developed illustrates the attitude to seed often found among farmers throughout the world.

In Bolenta, farmers now use the name *Himblil* to denote both the local variety and the variety Mr Negash developed based on it. More than 35 per cent of the villagers are now more interested in his *Himblil* variety than in the local one. This shows that the local farmers are not averse to trying new varieties as long as they believe these varieties are adapted to the local environment and will perform well. Releasing varieties developed by or in collaboration with local farmers, as in this case of *Himblil*, is therefore a promising approach to overcome this barrier and has the potential to increase yields while also contributing to the conservation of crop genetic diversity. By promoting benefit sharing and the protection of traditional knowledge, such projects also contribute to realising Farmers' Rights.

Figure 11.2 The barley variety Himblil developed by Mr Negash and his family
Source: Fetien Abay

Collaboration with Mekelle University

Researchers from Mekelle University started working with Mr Negash in 2002 and since then he and the other villagers have been motivated to pay more attention to varietal selection. The university involvement strengthened the existing practices and knowledge in the village through joint experimentation with local farmers. This partnership also benefited farmers outside the project area as the barley varieties developed were later introduced in other villages as well. Traditional knowledge has in this way been kept alive and supplemented through *in situ* conservation, active use and innovation, and the benefits from this are being shared through the collaboration with a research institution.

A follow-up project funded by the Norwegian Cooperation Programme for Development, Research and Higher Education was begun in 2007 and ran until the end of 2012. This project aimed at improving productivity and food security in Tigray through better access to quality seed. Decentralised participatory plant breeding methods were used to ensure that the resulting varieties would be adapted to local conditions, that they would be compatible with local farming systems and that they would satisfy farmer requirements. The project has shown how specifically adapted traditional varieties are used in variable environments and how farmers use genetic variation within crops to match variation in soil conditions between and within fields. The project also investigated the potential of barley as a high-quality alternative to teff for making *injera* and the opportunities for fair-trade export of *kolo* made by local women's groups.

As part of their collaboration with Mr Negash, Mekelle University included his varieties in its Participatory Varietal Selection programme along with formally released varieties. Through this programme, further evidence emerged regarding *Himblil*'s flexibility across various sites with growing seasons of varying duration, and its non-lodging traits were also confirmed (see Abay and Bjørnstad, 2009). Another indication of its promise is the fact that most of the farmers participating in the programme rated this *Himblil* very highly.

Participating in an exposure visit in 2009, Mr Negash was pleased to see that through the university's project his varieties of *Demhay* and *Himblil* had been introduced at 12 sites in Tigray and two sites in the central part of Ethiopia. He found particular reason to be proud of the *Himblil* and its excellent performance. However, his *Demhay* variety did not perform well on sites with less fertile soils and little rainfall; moreover, it proved susceptible to diseases and had a relatively low grain and straw yield. These results point up the potential weaknesses of local seed selection: such selection might result in lines that are excellently adapted to a very specific place under highly specific circumstances, without environmental variance being taken sufficiently into account. Pairing farmer knowledge and preferences with the scientific competence and selection methods of formally trained plant breeders and scientists, as in the case of the collaboration between Mekelle University and Tigrayan farmers, thus represents a very promising approach.

Farmer breeders are often more open to innovation and change than farmers who do not practise selection and breeding; moreover, they usually receive local recognition for their practices. Mr Negash's breeding efforts have benefited not only himself and his family but also their village. Through participation in the Mekelle University projects he has been both a recipient and a source of benefit sharing. And since Mr Negash started his barley breeding work, awareness regarding seed selection and varietal selection has increased in the community – one of the resulting benefits that also contributes to keeping the traditional knowledge alive. The complex seed flow networks documented by Abay *et al.* (2011) may in this way be mobilised for efficient dissemination of new varieties. In 2011, *Himblil* was distributed for seed multiplication among more than 5,000 farmers.

In the mentioned follow-up project, a key resource was a cross between *Himblil* and the low-land landrace *Saesa*. This cross created several new and intriguing trait combinations. Two of them were accepted for official release in February 2012. This also illustrates that farmers are often limited by a lack of genetic diversity. The diversity currently existing in their fields is much smaller than what can be created through deliberate selection and breeding.

Concluding remarks

The relationship between environmental variability and crop diversity found in Tigray is by no means unique to Ethiopia, so specifically adapted traditional varieties and traditional knowledge should also play an important role in other locations. In difficult environments, officially released varieties will often have a poor adoption rate because they are not sufficiently well-adapted to local conditions. Releasing varieties bred by farmers or with farmer participation based on local genetic material is likely to be more successful. Moreover, involving farmers in plant breeding and variety release together with scientists can serve to invigorate existing traditional knowledge and improve farmers' knowledge about varietal selection. By supporting farmer innovation and releasing farmer varieties it is possible to contribute to the maintenance of local crop diversity and traditional knowledge, as well as to the improvement of food security. In this way, important steps can be taken towards the realisation of Farmers' Rights.

Better knowledge about varietal selection becomes particularly important when the environmental conditions in an area are changing, for example due to climate change. Varieties that can adapt to the changing circumstances then become even more essential to food security. Traditional knowledge must be utilised if it is to be kept alive, and this example from Ethiopia has shown how such knowledge can evolve and adapt to change. It also illustrates how two elements of Farmers' Rights – the sharing of benefits from the utilisation of genetic resources, and the protection of traditional knowledge – can be promoted in a mutually supportive manner and that research institutions and scientific knowledge can contribute in this process.

The existence of traditional knowledge enables farmers like Kahsay Negash to serve as innovators and improve their livelihoods through the enhancement of traditional varieties. One of the lessons that can be learned here is that farmers are discerning users who will adopt only those varieties that fit their needs and are well-adapted to the local environment. Mr Negash's *Demhay* variety, although an improvement in terms of drought-resistance, failed to gain wider acceptance because of its longer growing period and preference for certain soil types whereas his *Himblil* variety has become quite popular due to its adaptability, good yields, tall stature and suitability for making *injera*. Through collaboration with university scientists, farmers can benefit from scientific knowledge and new technologies and methods, and this is an important part of benefit sharing.

In Tigray, traditions of farmer selection can still be seen, and traditional knowledge is conserved through practise – and these are ideal conditions for collaborative breeding projects between farmers and researchers.

Notes

1   When improved modern varieties are grown, they are usually of wheat and not barley.
2   This is also confirmed by Addis Abraha Hindeya's PhD dissertation, *Barley (Hordeum vulgare L.) for Food: Breeding for Improved Nutritional Quality of Local Foods in Northern Ethiopia*, submitted at the Norwegian University of Life Sciences in October 2012.

12 Capacity building and farmer empowerment in Mali

Mamadou Goïta, Modibo Goïta, Mohamed Coulibaly and Tone Winge

In the Mopti region of Mali, in the district of Douentza, the project 'Seeds of Survival' (SoS) has brought substantive results since it was initiated in 1994. This project focuses on the conservation of local, traditional varieties, the sustainable use of these varieties and the exchange, protection and promotion of the associated traditional knowledge.

As a result of project activities, the farmers involved have increased their production considerably and are now more aware of the advantages and importance of maintaining local crop diversity and the related traditional knowledge. One such farmer is Aly Ongoïba. In addition to yield increases, he has seen local varieties being brought back to use and that awareness about the importance of conserving the local plant genetic heritage has increased. Through benefit sharing as a result of increased diversity and improved maintenance of local varieties and traditional knowledge, the SoS project in Mali has made significant contributions to the realisation of Farmers' Rights in the country. Capacity building and farmer empowerment have been central in achieving this.

The Douentza project was developed as a partnership between USC Canada, a Canadian NGO and local partners. These partners include farmers' organisations at the local level, local governments, local associations and NGOs, and local administrative authorities. A central element of the project is the focus on capacity building, farmer involvement and cooperation between stakeholders to create long-lasting improvements. This is a strategy that has proven successful – even after the coup and rebel threat of spring 2012 caused the USC to shut down its offices in the country, local farmers continued programme activities in their fields and communities.

The Seeds of Survival approach

The SoS programme was created by USC Canada (originally 'Unitarian Service Committee of Canada') in 1989 in Ethiopia. USC was founded in 1945 and works within the area of rural development, with a particular focus on agricultural biodiversity and its importance for the livelihoods of farmers.[1]

The SoS programme was created to promote long-term food security for marginal farming communities in developing countries. It works to combine the knowledge of scientists regarding how to improve local crops, with the traditional knowledge of farmers. A main objective is to support communities involved in the maintenance of crop diversity by improving their capacities through participatory research and experimentation. Central to the programme is the belief that in order to achieve sustainable poverty reduction the management of this diversity should be decentralised and focused on improving food security and revenue generation for farmers. One way the SoS programme does this is by evaluating the existing plant genetic resources, thereby adding information to the local knowledge systems.

The SoS programme works with farming communities in many countries to create a platform for dialogue between farmers, NGO workers, scientists and government representatives and to facilitate farmer-to-farmer exchanges. Encouraging the conservation and enhancement of local crop varieties is important in the programme and promoting the creation of small community seed banks is a central part of this. The programme also promotes natural resource management systems that are adapted to diverse production systems and conditions and works to improve the ability of farmers to manage seed-supply systems and influence public policies. To educate and interest the younger generation, the programme promotes and facilitates the creation of school arboretums. All activities are conducted in partnership with local institutions – NGOs, farmers' groups, local authorities and research institutions.

In the SoS programme farmers are seen as experts, and knowledge-sharing is therefore a key component. Meetings and technical conferences where farmers from various countries share their experiences on different subjects with each other are organised regularly to encourage information sharing and training. At these meetings agricultural specialists also discuss their experiences from SoS with other practitioners within the area of food security and search for new ways to improve existing practices and apply SoS concepts to varying local conditions.

Through these meetings collaborative relationships are established between farmers and between farmers and scientists. As a result of these gatherings, participants have set up farm-based seed-saving programmes in at least 59 countries, and in this way biodiversity-based agricultural principles are spread around the world. The SoS model itself has also spread and evolved into a global programme with partners in countries such as Bangladesh, Bolivia, Burkina Faso, Cuba, East Timor, Ethiopia, Honduras, India, Indonesia, Lesotho, Mali, Nepal and Senegal.

The SoS approach has the potential to contribute greatly towards the realisation of Farmers' Rights under the Plant Treaty because it promotes benefit sharing in the form of improved access to seed, increased food security, information dissemination and sharing, and farmer empowerment. In addition, its focus on the role of farmers as custodians of agricultural biodiversity and on activities to promote the maintenance of crop genetic diversity and the related traditional knowledge, contributes to the implementation of Farmers' Rights.

Seeds of Survival in Mali

The SoS programme in Mali was inspired by the project in Ethiopia. The starting point for the programme in Douentza was the training of three USC staff members from Mali during the annual 'Seeds of Survival' training sessions in Ethiopia in 1993 and 1995. The SoS approach from Ethiopia was then adapted to the local context in Mali.

For many years after Mali held its first democratic elections in 1992, the country experienced relatively rapid economic growth and stability, and the democratic regime and decentralisation of power facilitated the undertaking of development projects. However, in March 2012, against a backdrop of unrest and rebel activity in the north of the country, a junta seized power in a coup, claiming that the government had not done enough to stop the northern rebels. Since then the rebels have seized control of northern Mali and declared independence, but this declaration has not been recognised internationally.[2]

Mali is one of the poorest countries in the world. It was ranked 175th out of 187 countries by the UNDP in its 2011 Human Development Index.[3] Agriculture, with its GDP share of 38.8 per cent,[4] is a very important sector; the majority of the population live in rural areas.

Because of the central role of agriculture in the economy, the government has traditionally paid attention to the sector. The agricultural sector has also been organised in a way that takes the dominance of small-scale family farms into account. With the passing of the *Loi d'orientation Agricole* (LOA) in 2006, the framework law that determines agricultural policy in Mali, an important step was taken towards organising the farming sector and giving it a prominent place on the development agenda. Because of the participatory approach used throughout the drafting process, this law was welcomed by the stakeholders.

The process was led by CNOP/Mali *(Coordination Nationale des Organisations Paysannes du Mali)*, which coordinates the various farmers' organisations in the country. This was the first time Malian farmers were given the responsibility to head a process of framing a major policy document. All consultations were led by farmers with the help of other civil society actors.

However, the legislation in Mali governing seed production and marketing is not particularly conducive to the realisation of Farmers' Rights: it does not recognise farm-based seed and allows only seed from registered varieties to be marketed. The marketing of farm-based seed from unregistered varieties is indirectly prohibited. Despite this, most farmers continue to use and exchange seed from such varieties because of their availability and adaptation to the local environment and culture. Many stakeholders in Mali therefore argue that what is now needed is a new law that can take into account the type of seed farmers actually need and use.

The seed regime in Mali is also influenced by the Economic Community of West African States (ECOWAS) and its efforts to harmonise the legislation of its member states. In addition, Mali has signed and ratified the Convention on Biological Diversity and the Plant Treaty, which recognise the rights of farmers and local communities and aim to protect their traditional knowledge related to biodiversity. These international agreements have been of great importance to the SoS programme in Mali in terms of motivation, and they have also influenced its activities.

The Douentza project

The SoS programme in Mali is implemented in the district of Douentza in the Mopti Region. Mopti is situated in the central part of the country, and Douentza is the region's northernmost district.[5] The district is divided into rural and urban 'communes' or local governments, with each commune composed of various villages. The total area of Douentza is about 23,481 square kilometres and the district had a population of around 247,794 people according to the 2009 census. Major ethnic groups in the district are the *Fulanis* (or *Peulhs*), which make up 43 per cent of the population and the *Dogons* with 30 per cent.

A rainy season lasting from June to September and a dry season lasting from October to May dominate the local climate of Douentza. Temperature may range from 10°C at night during December and January to 45°C in the daytime during April and May.

The district economy is based mainly on agriculture and animal husbandry. Millet production constitutes 85 per cent of the agricultural land used, with several different varieties grown. The local farmers also cultivate other crops, such as sorghum, rice, beans, peanuts and sesame.

The SoS project has partners in 18 villages, as well as in the town of Douentza. These sites can be divided into three categories based on the density of project activities: some villages have a high concentration of project activities, while others have a medium amount and the third category is made up of newly selected villages with fewer activities. The crops covered by the project can also be divided into three categories: cereals (like pearl millet, sorghum, maize, rice, *fonio* and wheat), leguminous crops (such as cowpeas, pigeon peas, groundnut, beans and Bambara nuts) and vegetables and garden crops (such as okra, calabash, hibiscus, squash, watermelon, eggplant, garlic, onion, pepper and hibiscus). Different groups work on these crops in terms of seed production, protection and promotion. These groups also work on facilitating seed exchange among communities.

Many different stakeholders are involved in the SoS project in Douentza, and they interact in order to facilitate implementation of the various activities. USC Canada is responsible for financial and technical support, with a technical team based in Douentza and a coordination team based in the national capital, Bamako. Other key stakeholders include the local government and administration, technical services at the local level, individual farmers and farmer organisations, schools, local associations and NGOs.

The SoS project in Douentza aims to meet the challenges that local farmers face with regard to seed. That means that farmers are deeply involved throughout the entire process, from the identification of activities to the evaluation of their implementation. In the planning stages, farmers and other stakeholders meet at the village level to discuss activities implemented during the previous year and identify key activities to be planned and the associated costs. They then negotiate with technical staff from the USC and agree on activities. Local committees are set up to manage each activity and the necessary resources. It is the members of these local committees who are responsible for implementation and who report back to the other members of their organisations and to the other stakeholders. Because farmers are the key players in the SoS project, the participating farmers have been trained in the implementation of all project activities. In addition, training sessions and exchange visits have been organised to promote institutional development.

Activities

The development of seed-supply systems focusing on conservation and sustainable use is central. This has been done, for example, by developing community-based infrastructure, such as gene banks. Four community gene banks have been set up in the region to preserve the local agricultural genetic resources. These gene banks promote diversity by supplying a broad range of varieties to farmers. In addition to halting the loss of genetic resources, these gene banks have also contributed to preserving the related traditional knowledge, as well as tools for seed conservation. The gene banks have helped many families maintain part of their family seed collection, and they have also spurred greater interest in the conservation of local genetic resources through *in situ* maintenance.

Moreover, these community gene banks contribute to climate change adaptation by making the existing agricultural biodiversity more widely available to the communities. In this way these communities have access to an important range of resources – varieties that have adapted to the local environment again and again, as well as the whole range of related knowledge accumulated in the course of generations. Altogether, the four community gene banks contain 758 samples, 72 species and 12 traditional tools related to seed storage.

Seed banks have also been established, and these have provided the farmers with increased seed security in a zone where poor rainfall sometimes means sowing must be done four to six times. Seed banks are critical when trying to establish secure seed systems and conservation programmes. They can help farmers when there is a rain deficit or a desert locust outbreak and farmers have difficulty in obtaining seeds from their own fields. Such local seed collections can solve the problem of seed availability and can serve as a reliable source of known varieties as opposed to any external seed that may be donated. In Douentza, the establishment of seed banks has contributed to greater solidarity among farmers, communities and villages.

Through the activity called 'Fields of Diversity', the project has worked to raise local agricultural biodiversity by involving schoolchildren and the general population together with scientists in the re-generation of varieties and species that have almost disappeared from the area. Many varieties are cultivated in the same fields, and their life-cycles are monitored closely in order to evaluate performance in relation to precise objectives. Varieties are then chosen to match the needs of the farmers. Through dialogue and exchange, this approach creates synergy between scientists and farmers, develops confidence in collaboration as an approach worth pursuing and maintains the dynamics among the various stakeholders involved in the maintenance of genetic resources. As a forum for knowledge sharing it has helped farmers to understand scientific concepts and scientists to understand and recognise farmers' knowledge. The approach has also given farmers the opportunity to map their knowledge and to reinforce the capacity of farmers' organisations.

Based on surveys, varieties in danger of disappearing have been targeted. In 1987 the national research institute Institut d'Economie Rurale (IER) carried out a seed survey in Douentza. The results, together with the results from a similar study conducted in 1993 by USC, created the basis for a comparative analysis of the loss of plant genetic resources for food and agriculture in the area. It was discovered that many crop varieties had disappeared or were in danger of doing do. Some of these varieties have been saved by the project activities – collected, multiplied *in situ* and shared in the communities. This work has contributed to the promotion of some of the local farmer varieties still existing. If no action had been taken to save them, these varieties would have been lost.

The SoS project has attracted national-level attention in Mali. In 2005 a 'seed caravan' was organised that caught the attention of decision-makers and highlighted the need to safeguard farm-based varieties. This caravan focused on the conservation and use of genetic resources and was mentioned on the national TV channels as well as on TV5, the international francophone television channel. The seed caravan was part of the SoS project component on promoting agricultural biodiversity. Together with seed fairs, it resulted in the collection of about 1,094 samples of cereals, vegetables and leguminous plants.

Facilitating the exchange of propagating material and the associated knowledge is an essential part of the SoS project. Important in this context are seed fairs, which play a vital part in the evaluation of existing plant diversity in farmers' fields. Participating at the fairs are experienced farmers with a wide range of skills related to the protection of plant diversity. Propagating material and related information and knowledge are shared without constraints.

Another project component focusing on the exchange of seeds and knowledge is the 'stock exchange'. This event is normally held before the rainy season starts and enables seed producers of farm-based seed and local seed buyers to meet. It allows farmers looking for specific varieties to access the seeds they need. Because it is organised right before the rainy season begins, this event represents an alternative way to access seeds that also allows the exchange of related knowledge. The 'stock exchange' can therefore be seen as an additional means of promoting plant diversity, as well as serving as an opportunity to give credit to seed producers for their work.

The component called 'environmental follow-up' has also been central. The goal has been to sensitise local communities to the negative impacts some of their actions might have on the environment, and the approach has helped them to understand more about the challenge of environmental management in fragile areas. Environmental education with a focus on sustainable management of biodiversity has been promoted in the primary schools, including the creation of school arboretums. These can be a very relevant and useful teaching tool because they give the pupils practical training in relation to environmental issues. In addition to scientific knowledge, it is important to teach the schoolchildren local knowledge about the environment they live in. This can be done by arranging for the village elders to share their knowledge and experience with the children, which creates intergenerational dialogue in line with the Malian way of safeguarding knowledge orally and contributes to the maintenance of local practices and traditional knowledge.

Together these activities constitute an approach that contributes to the realisation of Farmers' Rights in many ways: to benefit sharing and the protection of traditional knowledge in particular.

Practical experiences of SoS: Aly Ongoïba

Aly Ongoïba lives in the village of Petaka with his family. Petaka is one of the partner villages of the SoS project and lies about 15 kilometres from the town of Douentza. Mr Ongoïba and his wife, their five children and his mother live together in the family compound. Mr Ongoïba has been involved in many of the activities related to genetic resources management that have been initiated as part of the SoS project. His story is similar to that of many other farmer families living in Douentza. He is proud of how his life has changed as a result of the Seeds of Survival programme. As his experiences show, the project has indeed brought about changes in his life as well as in the lives of his family and his fellow villagers.

On his farm of about 6 hectares Mr Ongoïba mainly cultivates cereals, as well as some vegetables. He says that the village has always cared about their local varieties because agriculture is so important in the area and access to seeds is vital. They also consider their food-related traditions to be an important part of their culture: they value being able to eat tasty and healthy food in line with their traditions and possessing knowledge about local varieties and seeds. In addition, economic concerns also make the villagers prefer local varieties. Taste, economy and season all play a part in deciding which varieties to grow at any given time.

The village started their collaboration with USC Canada in 2000, but it was only in 2005 that they really started to focus on seed issues. Mr Ongoïba and 24 other villagers are now working on the conservation of local varieties and seed production. This group also studies how the different varieties perform in the soils in the area and tries to find new uses for their produce.

Mr Ongoïba feels that since the beginning of their partnership with USC the village has achieved many things. Local varieties that had disappeared because they were not being cultivated by farmers have now been restored. More than seven local varieties that were about to disappear from the region have been re-introduced to the area. This was done by collecting information from households, and eventually the village group found and collected seeds of the varieties in question. These are now being maintained in the area as well as being stored in seed banks. One of the varieties the village group has rehabilitated is a local variety of millet known as *Nioudaou*. This variety is regarded by Mr Ongoïba as very important and is now being cultivated in many villages. Mr Ongoïba thinks it is important that project activities focus on regeneration of and research on local plants because it allows the village to maintain the landscape and make it suitable for farming, cattle breeding and other activities.

As part of their SoS involvement, villagers also take part in training and exchange programmes. Further, they have organised gardening projects where, for example, women's groups cultivate vegetables for income generation and the children receive environmental instruction at school.

Petaka has its own seed bank now that contains seven varieties of sorghum, six bean varieties, about ten varieties of hibiscus and many varieties of millet and sesame. Mr Ongoïba is a member of the management group that runs this seed bank. He underlined that the USC Canada partnership has helped them to 'understand most of the challenges on seed issues'.

Mr Ongoïba also mentioned how project-related activities and techniques have allowed his family to increase production by almost 30 per cent, which he considers to be substantial. Many farmers who were reluctant in the beginning and who thought that local varieties could not compete with the modern varieties from the research sector have now joined the group.

By visiting other farmers, Mr Ongoïba has gained new knowledge that has allowed his family to improve their production techniques, and they now have more money. This is important to him even if he still feels he needs to do more to better his living conditions. Mr Ongoïba feels he is more aware of issues relating to food sovereignty than he used to because of the project, and he now works with other people to promote resource control at the local level. Through his participation in the SoS project and meeting scientists and other farmers he has realised that he too 'can be a scientist in my field', and this has added value to his work. He is able to select seeds by observation in his own field as well as in others and feels proud of this.

In addition to the improved production, Mr Ongoïba says that participation in the project has improved his reading and writing skills in the local language. Not only can he now read and write himself, he also teaches others.

Mr Ongoïba would like to call upon all farmers in Mali to join in the conservation and promotion of local varieties and to use them in their fields. This is his message:

> Agriculture is our work and we need to improve it, but also ensure that it is sustainable. Once you don't have control over your seed, you're lost. Let's then work together so that the decision-makers respect our right to use our seeds, and to be recognized as scientists too.

Achievements

Since its inception, the SoS project in Douentza has had an impact on some 32,300 people or about one third of the total population in the project area. During the first project phase about 1,514 households were involved, but this number has since increased to 6,475 or 23 per cent of the total number of households in the participating villages. Because of the changes in their seed practice and their participation in project activities, the livelihoods of these families have improved.

A comparative study conducted in Douentza and Fana (in Koulikoro region in Mali) showed that 95 per cent of the varieties cultivated by the farmers in Douentza were local varieties, whereas in Fana the corresponding figure was only 54 per cent. A similar study showed that in Podor in Senegal, 51 per cent of the varieties were imported. These differences between Douentza and similar areas are significant. It can be argued that the high use of local varieties in Douentza is a result of the activities of the SoS project.

The SoS project has built bridges between farmers and scientists and their occasionally differing perspectives and has created the foundations for a concerted approach towards conservation and use of plant genetic resources, at both household and community level. In this way it has contributed to the development of new types of collaboration between scientists and farmers. Central here is the fact that farmers are becoming increasingly confident of their own knowledge and practices and their scientific value. The SoS project has also contributed to strengthening the organisational capacity of farmer groups. Capacity building is a key component of project activities. One indicator of the importance of this work, and its success, was shown when a group of farmer representatives were able to take over coordination of programme activities when the USC office in Douentza was closed down in April 2012 due to the unrest.[5]

One of the main reasons for the successful achievements of the SoS project in Douentza is the strong commitment from the local communities, as well as from the development agents and scientists who worked with the farmers. The farmers and their organisations in this very vulnerable area were motivated to work on seed issues, and the sensitisation activities conducted early in the process provided the communities with a better understanding of what was expected from each group of actors. It was also important that studies conducted prior to the initiation of the project enabled a wider understanding of the region, the context and the key players with regard to genetic resources. Capacity-building activities were crucial to this success because they set the basis for working on genetic resources issues and seed initiatives in particular.

Achieving substantial progress would have been difficult without the collaborative relationship created among key stakeholders, such as individual farmers and their organisations, public technical services, local administration, traditional administration, local governments, NGOs and scientists. Entrusting responsibilities to farmers and their organisations was an important part of this. It was also central to the success of the project that the various project activities complemented each other and brought on board a range of interests to be taken into account. In addition, the financial, technical and material support from USC allowed the groups to move beyond what the circumstances would otherwise have allowed them to.

Challenges and lessons

In the beginning, the project faced various problems. Many local stakeholders were highly sceptical to the approach and found it hard to believe in the plans. In fact, the first challenge was to overcome the resistance of scientists who were not accustomed to working with farmers as equal partners. Setting up communication between the different stakeholders was therefore a first key step.

The second challenge was connected to local resistance to traditional varieties. Many people found it hard to believe that farmers' varieties could have the same potential in terms of yield growth as the hybrid varieties from research institutions.

There was some resistance among farmers to working with local genetic material, as they considered this material to be old-fashioned for contexts of drought and with no real performance. One team member of the USC staff noted how they were initially also met with scepticism when they presented their plans to focus on the conservation and use of traditional varieties to scientists. One scientist from the regional IER centre in Mopti had claimed that they would have to be crazy to use traditional varieties if they wanted to improve yields – but that same scientist has now joined the project. The results of the project have shown that local material indeed has potential and can be very useful in plant genetic resources development.

The weak organisational level of most of the farmer groups at the beginning of the project also presented an obstacle. Initiatives like the SoS project cannot succeed without strong farmer groups, so the project invested in training sessions, farmer-to-farmer exchanges and other capacity-building activities. In part, the weak policy commitments of the various farmer groups and farmers' organisations on issues related to genetic resources can be seen as related to this weak organisational structure.

Some problems and challenges have concerned climate change and how to deal with different changes in practices and policies, while other challenges have been related to the extreme poverty of the vast majority of the population in the region and stakeholders' generally weak policy knowledge regarding their rights. These problems have been taken into account in the project. They still exist to some extent, but much has been done to mitigate them.

A major challenge in the future will be to scale up the various activities, regionally and nationally, so that the project may have a critical impact on national policies. For this to happen, decisions will have to be made with regard to strategy. In part, the ability of the project to scale up will depend on the types of partnerships USC develops with other organisations and networks at local, regional, national and international levels.

Concluding remarks

As this story from Mali has shown, it is possible to build bridges between farmers and scientists and to bring their occasionally differing perspectives closer together through collaboration and knowledge exchange. When farmers are integrated into development projects as equal partners with expert knowledge of their own, their needs and experiences can more easily be taken as the point of departure for activities and that in turn increases the chances of success and sustainability.

Capacity building for farmers and farmers' groups can be central in this context, to ensure that the farmers can be involved in and take control of the implementation of project activities. The SoS programme in Mali illustrates how capacity building can lead to farmer empowerment and successful activities like seed surveys, community gene banks, seed banks, seed caravans and environmental education. The farmers have become more confident of their own knowledge and traditional practices and their scientific value. In addition, the organisational capacity of the local farmer groups and farmer organisations has been strengthened.

Dissemination of information, knowledge exchange and capacity building are all important benefits. Moreover, as this story from Douentza district shows, this type of benefit sharing can lead to further practical benefits in the form of increased crop production and improved livelihoods. These benefits have been based on the utilisation of local genetic material. The results have convinced both scientists and farmers originally sceptical to the value of such resources that local genetic material is worth using and that the related traditional knowledge must be maintained and utilised to achieve the best results. The SoS programme in Mali has therefore contributed to the realisation of Farmers' Rights by promoting benefit sharing.

Notes

1   See the USC website for more information www.usc-canada.org/who-we-are/
2   For more information about the situation in Mali, see www.bbc.co.uk/news/world-africa-19327916 and www.bbc.co.uk/news/world-africa-17582909
3   Human Development Index 2011 www.hdr.undp.org/en/statistics/
4   CIA – The World Factbook: https://www.cia.gov/library/publications/the-world-factbook/geos/ml.html
5   When the Tuareg separatist group declared Northern Mali an independent state in April 2012, they claimed Douentza as part of their territory. As a result, the USC office in Douentza closed down but programme activities continued. For more information see the USC website www.usc-canada.org/2012/04/16/mali-turmoil-affects-usc-programming/ and www.usc-canada.org/2012/08/30/access-to-local-seed-in-time-of-turmoil/

13 The Hiroshima Agricultural Gene Bank

Re-introducing local varieties, maintaining traditional knowledge

Yoshiaki Nishikawa and Tone Winge

The modernisation of agriculture has brought great changes to the farming societies in rural Japan. Until the 1970s, most Japanese farmers saved their own seed, especially in the case of vegetable crops. However, after the arrival of hybrid varieties many farmers started to buy seed from seed shops. This negatively affected the *in situ* on-farm conservation of plant genetic diversity because it meant that fewer farmers were maintaining traditional varieties.

With regard to rice, the number of varieties cultivated has been drastically reduced, both nationally and locally. The development in Hiroshima Prefecture illustrates this trend. In 1995, there were only 11 rice varieties on the prefectural government's list of recommended varieties, whereas more than 500 varieties had been recorded when the first agricultural research station was established in 1894 (Hiroshima Prefecture Agricultural Gene Bank, 1995). As the varieties gradually disappeared from the countryside, the farmers lost access to the wide range of traits they contained; moreover, the related traditional knowledge went out of use and was in danger of vanishing.

More recently, however, along with the increased international recognition of the importance of conserving and sustainably using plant genetic resources, various initiatives have been taken to ensure that Japan's agricultural biodiversity is maintained and remains accessible to Japanese farmers. Such initiatives can contribute to the realisation of the various elements of Farmers' Rights.

This chapter tells the story of how an agricultural gene bank established by an independent research foundation with the assistance of the local government of Hiroshima is contributing to the maintenance of traditional varieties and their re-introduction to local farmers. The facilitation and technical input offered by the gene bank has made it possible for farmers to take up their old practices of seed saving and breeding. Through the gene-bank initiatives, not only the varieties themselves but also the related local traditional knowledge, including knowledge about seed harvesting, have been brought back and traditional practices and techniques are again being used. Keeping such knowledge alive is a very effective way of protecting it from extinction. The Hiroshima Agricultural Gene Bank has therefore contributed considerably to the realisation of Farmers' Rights.

Conservation of plant genetic resources in Japan

The importance of conserving and utilising what still remains of the country's plant genetic diversity and traditional knowledge is now widely recognised in Japan, and various national institutions have been established for this purpose. The key national research institution with respect to this work is the National Institute of Agrobiological Sciences (NIAS). The NIAS Gene Bank, established in 1985, coordinates efforts to conserve plant genetic resources for food and agriculture. It currently contains more than 248,000 accessions. Although mainly concerned with maintaining and providing seed, for some of the accessions collected within Japan the gene bank also possesses documentation of traditional knowledge; for example, related to processing methods and use in rituals.

Until recently, farmers were major contributors to the conservation of Japan's crop genetic diversity and the protection of the related traditional knowledge. Today it is mainly organic farmers who play an important role in this context. According to a 2010 survey conducted by the Japan Organic Agriculture Association, 58.7 per cent of organic farmers were multiplying vegetable seeds by themselves; the corresponding figure for tubers was 66.9 per cent and for pulses 62.3 per cent (Japan Organic Agriculture Association, 2010). This shows that these practices are still in use, at least among some farmers. They grow mostly traditional varieties, but some cultivate non-hybrid modern varieties as well. The survey also showed that, for the farmers in question, the increased independence of their farms, the improved possibilities for adapting the material to farm-specific conditions, the lower costs and the opportunities the practice brings for growing local varieties and traditional crops were seen as the most important benefits of seed-saving (Japan Organic Agriculture Association, 2010).

Access to traditional knowledge is often central to succeeding with the cultivation of local traditional varieties, so such knowledge contributes to the maintenance and enhancement of agricultural biodiversity. In Japan, there is a considerable amount of traditional knowledge related to what foods to prepare for various special occasions, such as the New Year celebrations and the ceremony for welcoming ancestral spirits.

Both the private sector and the state face limitations with regard to their ability to maintain the remaining plant genetic diversity and traditional knowledge in Japan, as the 'westernisation' of the diet has drastically changed consumer behaviour. For the farmers, who are the key actors in the management of local varieties, lack of technical knowledge and the absence of networks to other farmers and other stakeholders make it difficult for them to ensure that all such varieties and the associated traditional knowledge are maintained.

It is therefore important for the various actors in Japan to join together in establishing an efficient system for the *in situ* management of landraces and other traditional varieties as well as for the maintenance of associated traditional knowledge. There have been some examples of successful networking efforts and achievements, and non-profit organisations and institutions like the Hiroshima Agricultural Gene Bank have been particularly central.

Hiroshima Agricultural Gene Bank

The Hiroshima Agricultural Gene Bank was established in December 1989. Its history illustrates how the structure and funds of earlier institutions can be successfully utilised for establishing new institutions. Its predecessor was the Hiroshima Reclamation Cooperative, an important institution established during the Second World War for agricultural development and resettlement in remote rural areas of the prefecture. In the 1980s, when the Japanese government increasingly started to emphasise industrialisation and rapid economic growth, the cooperative lost its mandate. The idea of establishing a gene bank was born out of the wish to utilise the cooperative's funds in a good way when it was disbanded.

It was decided that the primary purpose of this new institution should be to collect and conserve genetic resources as well as to create a system to facilitate the improvement and utilisation of crop varieties and in that way contribute to rural development. And as the gene bank was meant to focus on local varieties indigenous to the various villages of Hiroshima Prefecture, the prefecture council decided that the prefecture itself, rather than the state, should establish and manage the gene bank. To ensure its independence the gene bank was set up as an autonomous foundation.

Various stakeholders collaborated in the effort to collect local crop genetic resources in immediate danger of disappearing, with local farmers, plant breeders and the Hiroshima Prefectural Technology Research Institute playing a main role. In addition, Hiroshima Prefectural University contributed with both information and accessions. The participation of former extension workers made it possible to collect and document traditional knowledge, not least related to the cultivation of the collected varieties and their place in the farmers' daily lives. For example, certain traditional vegetables were used to treat headaches and digestive problems but the knowledge regarding such uses would be found only in small areas and among members of the older generation. When the former extension workers collected genetic material in the local villages, they took care to record this knowledge.

A central activity of the gene bank has been a 'seed loan' system, which provides farmers with seeds from collected varieties (see also Nishikawa, 2001). Through this system farmers are able to obtain, either directly from the gene bank or through a local extension officer, seeds of varieties they wish to try out that are no longer available through the standard market mechanisms. Farmers who receive seed through this system are obliged to return the same amount of seed after harvest, together with a report on how the variety performed. Neither the national gene banks nor other institutions have implemented similar systems: the Hiroshima Agricultural Gene Bank is currently the only institution in Japan that gives farmers access to seed in this way. Since 2001 the gene bank has been able to provide about 200 samples every year. In addition, the gene bank offers advice on how the different varieties can be used as well as information regarding their original cultivation area and cultivation methods, based on traditional knowledge.

Since its establishment, the Hiroshima Agricultural Gene Bank has also been providing plant genetic resources to the Hiroshima Prefectural Technology Research Institute as well as to plant breeders and Hiroshima Prefectural University.

Initially, the gene bank did not actively undertake seed collecting. The focus was more on received seed and on introduced material than on local material. However, when Ataru Okimori, a vegetable specialist who had previously been a prefecture official, joined the gene bank, he suggested that it should appoint individuals responsible for actively collecting vegetable seed from the various regions in the prefecture. This marked the start of the 'Farm-to-Farm Search for Genetic Resources' of the 'Three-Year Strategy of Hiroshima Prefecture' from 1992. Then, in 2009, the gene bank launched a new project on what have been termed 'treasure vegetables' and the use of such varieties to re-vitalise local farming.

The farm-to-farm search campaign

The farm-to-farm search campaign aimed at locating endangered varieties that were still cultivated in the prefecture and determining how they might be of use today. This search resulted in the collection of altogether 387 seed samples as well as information about the location of 160 old fruit trees. Traditional knowledge was also documented through communication with farmers and other local residents. Such knowledge includes information about where in the field certain varieties should be grown as well as information on processing and cooking.

The farm-to-farm search campaign was initiated in 1992 because of fears that agricultural modernisation was posing a serious threat to local varieties and the associated traditional knowledge. Mr Okimori felt that efforts to collect and conserve local varieties were urgently needed, as many varieties were in danger of disappearing for good. He therefore proposed a plan for the collection of such crop genetic resources. One of the unique features of this plan was the way it utilised the knowledge of retired extension officers in each of the prefecture's ten regions, giving them the responsibility for conducting the search in their own region. These chosen retirees were familiar with the traditional farming practices of the various regions and thus represented a valuable resource for the project. Another central part of the campaign's strategy involved collaboration between these former extension officers and local organisations, such as agricultural cooperatives and clubs for the elderly.

Collecting seeds was not easy in the beginning. One of the obstacles was the belief held by many farmers regarding the need to keep their seed heritage on the farm and how it would be wrong of them to allow some of it to be taken away. For example, one farmer firmly resisted giving away his seeds because his father had told him never to take the seeds away from their farm.

Gradually, however, the collecting process became easier as more and more farmers offered their seed – even those who had initially refused. One of these farmers cultivated a variety of cucumber called *Aodai*. He had cultivated this variety since his wife brought it from her hometown of Fukuyama in the south-eastern part of Hiroshima Prefecture, and he had been committed to the idea of never sharing his own seed with other farmers or families. This attitude to seed sharing was something the search campaign sometimes encountered among farmers who bred their own varieties, but in many cases the farmers they met would at least share their seed with neighbours and relatives. However, when this particular farmer read in a newspaper that 'his' cucumber variety was in danger of disappearing, he decided to offer seed to the gene bank. This shows that even farmers who are inherently sceptical about giving away their seed due to cultural traditions may want to share when they learn about the threats to the varieties they cherish. This example also illustrates the important part the media can play in disseminating information about the importance of maintaining crop diversity.

The farm-to-farm search campaign lasted three years until 1995, and the 387 samples that were collected came from altogether 130 species. The sample collection consists of 12 rice samples, 16 wheat and barley samples, 44 samples of pulses, 47 samples from millet and fibre crops, five samples from pasture and feed crops, two fruit samples, 250 vegetable samples and 11 samples from flowers and ornamentals. Among them are valuable varieties that had almost disappeared. After the passport data had been recorded and the germination ability tested, the seeds were dried and stored in airtight containers. A central principle of this collection is the obligation to return seeds to the donors if they so request. A list was also prepared of varieties that had disappeared from farmers' fields but whose prior cultivation could be confirmed (Nishikawa, 2001).

The Treasure Vegetables Project

In 2009, the government of Hiroshima Prefecture introduced a policy for utilising local vegetables as part of its rural development policy and entrusted the gene bank with launching a new project, 'The Project for Vitalizing Local Farming by Means of Treasure Vegetables in Hiroshima' (hereinafter: the Treasure Vegetables Project). The aim was to select useful vegetable varieties indigenous to Hiroshima Prefecture that in some way were especially tasty, rare or associated with unique preparation methods and therefore worthy of being classified as 'treasure vegetables' and to re-introduce and utilise these valuable plant genetic resources by providing seed to farmers and disseminating information to various vegetable sellers (Nishikawa, 2011).

According to Tatsuaki Funakoshi, a gene-bank curator and vegetable seed specialist, it was decided that the project should examine the basic characters of 1,500 accessions stored in the gene bank, with a view to selecting 150 varieties as prospective 'treasure vegetables' based on their properties. The project would then encourage farmers and other vegetable producers to cultivate these and make them known to distributors and consumers. The gene bank is responsible for both the characterisation of accessions and the distribution of varieties. By fostering greater awareness in local farming societies of the importance of traditional varieties and the associated traditional knowledge and thereby contributing to the protection of this knowledge, this project is promoting the realisation of Farmers' Rights.

When the project was initiated, the number of farmers in the area was decreasing in line with the general ageing of the population. About 200 agricultural production groups engaged in the production, processing and distribution of produce were believed to be in danger of going bankrupt in the Hiroshima region. It was hoped that through the work of the gene bank, the cultivation of 'treasure vegetables' would catch the interest of younger generations and help to vitalise the region.

Three key elements

The Treasure Vegetables Project consists of three parts. The first involves the characterisation of each vegetable variety in the gene bank. This process was started in 2009 and will continue until the end of 2012. The second part of the project is the selection and multiplication of 'treasure vegetables'. In 2010, out of 50 finalists with excellent properties, five varieties were chosen: the *Aodai* cucumber, *Kan-on* leek, *Yaga chisha* lettuce, *Kawauchi* spinach and *Sasaki-Sangatsu Kodaikon* radish. Since then, five more varieties have been chosen each year.

A central and quite unique element of the selection process was the integration of tasting sessions in connection with the characterisation of varieties. Together with farmers and selected consumers, the gene-bank staff most familiar with many varieties tasted each variety, usually in cooked and/or pickled form, to examine its characteristics. Historical aspects were also taken into consideration and because part of the objective was to re-invent tradition and find new uses for varieties that had gone out of use, other preparation methods than those traditionally associated with a local variety were also tried. Sensory analysis was an important tool in this process.

All the chosen varieties were among the local varieties that had been collected after the establishment of the gene bank, and they all received very good marks during the evaluation process. *Aodai* literally translates as 'large-and-lush green cucumber'. It has a sweet taste and is particularly suited to be cooked for use in salads and for pickling. The *Kan-on* leek (from Kan-on district) has fleshy and soft leaves, is highly resistant to cold and disease and particularly suited for use in casseroles, in *Okonomiyaki* pancakes (Hiroshima-style pancake with vegetable filling) and for flavour in noodle dishes. *Yaga Chisha* is a lettuce from Yaga district; it is purple-red and yellow-green, has fleshy soft leaves and is slightly bitter. It is well-suited for use in roast meat salads, as garnishing served with sliced raw fish and salad dressed with miso. *Kawauchi* spinach comes from Kawauchi district, and *Sasaki-Sangatsu Kodaikon* is a giant radish that is highly cold-resistant, has fleshy roots, a mild and sweet flavour and is well-suited to vinegar-seasoned dishes, tempura, pickling and some soups.

The multiplication process for treasure-vegetable seed takes place without the use of fertiliser, and cultivating conditions differ from variety to variety. As part of the third part of the project, these varieties are now being promoted in order to increase their cultivation and seed production is also utilised for this purpose. To create interest in the local varieties and get interested growers involved in the selection process, presentations are held at the gene-bank farm twice a year. Between 20 and 30 varieties can now be seen growing there. In connection with these presentations, staff members talk with visitors interested in a particular variety about cultivation. As a result of this promotional work, restaurant managers, local shop-owners and others have now become interested in introducing the traditional varieties, and some have developed new recipes to promote value addition and consumption.

Results and findings

Experience with the Treasure Vegetables Project has shown that it is very important to provide farmers with detailed information on the properties and cultivation methods of the various local vegetable varieties as such knowledge is crucial to their motivation to continue cultivation. The traditional knowledge associated with these local varieties was forgotten when the varieties went out of use. It is therefore necessary to distribute detailed knowledge along with the seeds, as the methods for cultivating these local varieties tend to differ from those used in the cultivation of modern varieties.

For example, many of the local varieties have quite specific needs when it comes to the use of fertilisers and water, and the ideal production conditions must therefore be made known to all potential growers. Importantly, most of the traditional varieties do not respond to excess application of water and fertilisers as most modern varieties do. In addition, when farmers grow local varieties they must be mindful of weather conditions, as the different varieties may require specific conditions when it comes to soil quality and the timing of sowing. Traditional knowledge regarding these aspects, now backed up by the research results of the gene bank, has proved essential to those wishing to grow these vegetables.

Mr Funakoshi feels that the local vegetable varieties have the potential to gain substantial popularity because they are well-suited to organic farming (among other things). As these varieties have usually been grown with very little fertiliser, their seed therefore is better suited to organic farming than seed from modern varieties. He also thinks that the local varieties have the ability to adapt well to climate change.

As emphasised by Mr Funakoshi, when growing traditional local varieties, it is essential to match the right variety to the right location. Dissemination of traditional knowledge is crucial to succeed with this. Focusing on this in the extension work has proven a valuable strategy for increasing the use of these 'treasure vegetables' and other local varieties, and it is hoped, will ensure the stability and sustainability of the project.

The Society Committed to High Quality Crops: a farmer group in Sera

In Sera town in Hiroshima Prefecture, farmers have joined together in a seed-production group called 'Society Committed to High Quality Crops'. This group was established in 2001 under the guidance of Mr Funakoshi. The members grow vegetables, including 'treasure vegetables'.

Establishing the group

At the time when the 'Society Committed to High Quality Crops' was set up, there were 61 agricultural cooperatives in Sera; all consisting of farmers working together to take care of agricultural fields that were not efficiently utilised due to depopulation and the ageing of the farmer population. All of the involved farmers maintained their own varieties, for example of the local soya bean, and then exchanged the seed with each other and with neighbouring farmers. The 'Society

Figure 13.1 The group leader of the 'Society Committed to High Quality Crops' in front
of the group's vegetable field
Source: Hiroko Kubota

Committed to High Quality Crops' was established when out of these 61 groups, a
group that had specialised in traditional vegetables and another that had focused on
local cuisine got together and started to grow and multiply seeds from traditional
local vegetables. The group was provided with information and technical instruction
through Mr Funakoshi and received 'seed loans' from the gene bank.

As of late 2012, the group had 21 members. Each of these members has several
tasks, including producing vegetables, selling them and producing seeds. Group
meetings decide, by consensus, what varieties will be produced and by whom.
Membership is limited to residents of Sera but otherwise anyone from the town can
join the group if they are interested in cultivating traditional vegetables and sharing
the responsibility.

Since the group was established and word began to spread about its objectives and
activities, the number of members has grown and also young people have started to
join. One of the members explained that the new members have had many different
reasons for joining and their familiarity with the group's activities has varied. As
growing and multiplying seed from traditional varieties requires a substantial amount
of labour and is not very profitable, the group does not attract those merely interested
in making a profit. One goal of the group is to recruit members with noteworthy
farming skills, as that would improve the overall knowledge among the members and
benefit their activities and possibly also bring additional income.

The group has been quite successful in its activities. Despite the earlier trend of declining sales for traditional and local vegetable varieties, sales have been increasing.

Farmers' views on conservation and the challenges facing the group

As group members see it, the cultivation of local varieties is not just about conservation for its own sake but is linked to warm feelings that the producers have for vegetable cultivation. One of the original members of the group, Mr Kunifuji, underlined the joy and pride he feels about producing high-quality traditional local vegetables and the satisfaction that comes from seeing the vegetables grow as a result of his hard work. Mr Kunifuji also thinks that seed saving and seed production are among the factors that make farming fascinating and that when farmers do not engage in this work, farming becomes less interesting.

It is the women of the group who are in charge of the seed production. Because high-quality seed is essential, this work is regarded as very important. Two of the members, Ms Kato and Ms Kodama, emphasised the good taste qualities of the traditional varieties compared to modern hybrid varieties. They therefore think it is a good idea to re-introduce varieties that have disappeared.

Vegetable cultivation is also a source of income. Both Ms Kato and Ms Kodama underscored that one problem associated with cultivating traditional varieties is that it is almost impossible to make an acceptable living from cultivating only traditional varieties: they are inferior to the hybrid varieties when it comes to yields and pest tolerance. For this reason, they themselves try to cherish older tradition as well as earn a living by using both local and hybrid varieties.

Although the group has been quite successful, it is also facing some challenges. Cross-breeding is one of these challenges, and members want to receive technical training so they can learn to breed seeds without cross-breeding. The case of the *Aodai* cucumber, which the town of Sera would like to promote as its own local specialty, can illustrate the difficulties. According to one of the seed producers, several seeds harvested from one such cucumber may produce only one individual that possesses the desired traits because this variety cross-breeds so readily. It is not possible to see if a seed possesses the desired traits until a harvested seed bears fruit. The group therefore wants to learn how to avoid cross-breeding so they can grow more of this variety successfully.

The group members are also wondering how much of a difference their efforts to maintain traditional local vegetables have made and to what extent their work has had an effect on agriculture in the region. One barrier they are facing concerns attitudes and practices among young people. Members of the younger generations usually buy seed instead of saving and multiplying it themselves, and many regard vegetable cultivation as tiresome and un-profitable. Even the members' own children usually know little about the cultivation of local varieties: the difficulties associated with making a living from growing traditional vegetables is probably one reason for this. Although making such activities profitable is challenging in most rural areas of Japan, it is somewhat easier for farmers who grow varieties specifically adapted to their own local area to profit from their efforts. In addition to the benefits this type

Figure 13.2 A member of the 'Society Committed to High Quality Crops' explaining the
group's activities to visitors
Source: Hiroko Kubota

of adaption brings for cultivation, this is because there are still some consumers old
enough to remember how to use those local varieties and who appreciate having
access to them.

So far the challenges have not discouraged the 'Society Committed to High
Quality Crops' of Sera. The members find vegetable cultivation and seed production
inherently interesting. Their main goal is not to profit economically but to improve
the quality of life for themselves and for their community.

Concluding remarks

As this story from Hiroshima Prefecture in Japan shows, it is possible to achieve
considerable progress in realising Farmers' Rights when different stakeholders are
able to work together. In Japan, as in many other countries, farmers play a very
important role with regard to the maintenance of local varieties and the associated
traditional knowledge and here the exchange of seed and knowledge is central.
However, when local varieties stop being cultivated in an area and the associated
traditional knowledge falls into neglect, the farmers need support from other
involved actors, such as the local government and institutions working with
agricultural development and the conservation of plant genetic resources.

In the case of the agricultural gene bank in Hiroshima, a search campaign for local varieties and the identification of particularly interesting and useful vegetable varieties through the Treasure Vegetables Project, along with a system of seed loans and dissemination of information, has provided local vegetable growers with greater incentives and opportunities for cultivating local varieties. Significant factors have been the commitment and support of the local government – which shows the important role that such institutions can play in the realisation of Farmers' Rights.

Securing the involvement of retired extension workers proved central to the collection and documentation of local varieties and traditional knowledge, as these persons were knowledgeable about local practices and where the local varieties might still be grown. The mutual trust between these retired extension workers and local farmers also made it easier to get the farmers interested in the project and willing to participate. From this, other institutions seeking to implement similar projects can learn a useful lesson about the importance of utilising the knowledge and networks of experienced field workers.

However, collecting efforts sometimes encountered resistance among the farmers to seed sharing. For these farmers, the cultural traditions and what they had been taught about the importance of protecting the plant heritage bequeathed to them by their forefathers meant that they were reluctant to give away seed. However, the case of the Hiroshima Agricultural Gene Bank has shown that such obstacles can be overcome through increasing awareness about the threats to agricultural diversity.

The experiences of the agricultural gene bank of Hiroshima and its projects also highlight the importance of disseminating traditional knowledge along with seed, to enable successful cultivation of local varieties among new growers. Because the gene bank has collected and verified such knowledge, it is in a position to teach interested farmers how to maintain the various vegetable varieties and to match varieties to the right locations as well as to disseminate information about processing and preparation methods. Through such re-introduction and dissemination, this knowledge is protected and being kept alive.

The Treasure Vegetables Project also shows that it can be central to focus on traditional varieties with traits that will be appreciated by consumers and to create a market for traditional varieties through promotional activities in order to succeed with the maintenance of agricultural diversity.

The farmer group from the town of Sera is among those who have benefited from the activities and support of the gene bank. Members of this group have received seed and assistance, have been quite successful in their seed production efforts – and they are proud of and satisfied with their work. All the same, because of the difficulties associated with cross-breeding they are looking for further technical assistance and training. This underlines the need for close collaboration between institutions such as the gene bank and those whose work ensures that agricultural diversity is maintained *in situ* from day to day: the farmers themselves.

This story also bears witness to the deep and heartfelt enthusiasm many growers of local vegetable varieties feel in connection with their work and the satisfaction they derive from successfully saving and multiplying seed – as well as the severe blow to the conservation of these resources that their loss would represent. One barrier to safeguarding the local varieties and associated traditional knowledge in Hiroshima lies in the fact that the farming population is ageing, and there has been little interest among the younger generations in maintaining local vegetable varieties because this work can be time-consuming and difficult. However, the fact that some young people have joined the farmers' group in Sera as well as the group's impressive sales results and the growing interest in local varieties created by the Treasure Vegetables Project, all give reason to hope that this trend may be turned around so that the concept of Farmers' Rights as defined in the Plant Treaty will be realised to an even greater extent.

Part V

Success stories from farmers' participation in decision-making

14 Advocacy for Farmers' Rights in Nepal

Tone Winge, Kamalesh Adhikari and Regine Andersen

When Nepal negotiated with the World Trade Organization (WTO) from 1998 to 2003 to become a member, pressure was exerted on the Nepalese negotiators to introduce new legislation on plant breeders' rights. The negotiators from Nepal were told that in order to be admitted to the WTO, Nepal would have to join the International Union for the Protection of New Varieties of Plants (UPOV). UPOV membership would necessitate legislation with strong protection of plant breeders' rights, at the cost of Farmers' Rights related to seed and propagating material.

The National Alliance for Food Security (NAFOS), an umbrella organisation of NGOs in Nepal, involved itself in this process. They contacted farmers' organisations and facilitated workshops with farmer groups to develop a position regarding the issue of UPOV membership. They also consulted business actors in order to ensure that their position would be beneficial to Nepal on the whole. Against this background, NAFOS organised a campaign in Nepal and supported the Nepalese delegation during the negotiations. The result was that Nepal rejected UPOV membership and agreed to develop a *sui generis* system for plant variety protection, more suitable to the needs of the country, by the end of 2005.[1] Nepal became a member of the WTO on 23 April 2004, and the national Ministry of Agriculture and Cooperatives has drafted a law on Plant Variety Protection and Farmers' Rights that recognises the rights of both breeders and farmers.

As the advocacy campaign proved very successful, it stands as a good example of how farmers and their organisations can have a say in decision-making at the national level.[2]

Agriculture and seed in Nepal

Nepal is one of the poorest countries in the world.[3] Since some 80 per cent of the population live in the countryside,[4] rural poverty is widespread. Agriculture is characterised by small and fragmented subsistence farming: more than 75 per cent of the land holdings are less than 1 hectare, with 47 per cent even less than 0.5 hectare (IFPRI, 2010). Moreover, domestic agricultural production has not kept up with the population growth. Ever since the 1960s, population growth has outpaced the growth in cereal production (IFPRI, 2010).

Seed-saving and seed-sharing practices are essential in Nepalese agriculture, and there is a high degree of seed interdependency among farmers. One reason is that most farmers cannot afford to buy seed; as much as 90 per cent of seed transactions take place on a farmer-to-farmer basis (Adhikari, 2008).

As to the formal seed sector, there are very few companies producing and selling seed in Nepal (10 as of 2010) and none of them are multinational. However, seed from multinational companies, mostly from India and other Asian countries, reach seed users in Nepal through formal and informal trade. The National Seed Company Ltd is the main supplier of seed from modern, improved varieties of the major cereals and vegetables (Gautam, 2008). The Seed Act of 1988 and its regulations govern seed certification, quality and marketing rules, and the Seed Quality Control Centre of the Ministry of Agriculture and Cooperatives is responsible for the certification and release of varieties.

According to Durga Adhikari, a representative of the Seed Entrepreneurs Association of Nepal, the Nepalese government is, unlike the Indian government,[5] not doing enough to support the seed sector. Viewing the state investment and university research as too limited, Mr Adhikari feels that seed imports are a threat to the domestic seed sector. In particular, Nepal's open border with India has posed a challenge to the development of the domestic seed sector and has led to unauthorised seed trade. In his opinion, there is a very real need to promote the domestic private seed and breeding sector. However, the private sector as such is not opposed to the recognition and protection of Farmers' Rights. Many of its central players support the idea of including both plant breeders' rights and Farmers' Rights in the new law, and this attitude might have made it easier for the authorities to say no to UPOV membership.

Nepal is a country very rich in biodiversity. Its agricultural biodiversity is especially central to poor households in rural areas. According to Madhusudan P. Upadhyay, a Senior Scientist at Nepal Agricultural Research Council (NARC), there exist some 2,000 local rice varieties and more than 600 varieties of finger millet in Nepal. However, genetic erosion has already resulted in quite substantial losses: Mr Upadhyay fears that 50 per cent of Nepal's crop varieties have disappeared, and environmental change and loss of local varieties due to climate change are increasingly becoming a problem. With the climate becoming warmer in many parts of the country, crops are climbing to higher altitudes and many varieties are disappearing – and with them, the associated traditional knowledge and related skills (see also Gautam, 2008). This development makes it increasingly important to maintain agricultural biodiversity so as to ensure future adaptability to the changing environment. The practice of seed saving and sharing is central to halt the loss of crop varieties and ensure that the associated knowledge is kept alive.

Effects of plant breeders' rights on Farmers' Rights

Plant breeders' rights relate to new varieties of plants, not to traditional varieties maintained and developed by farmers. Nevertheless, depending on their coverage and scope, plant breeders' rights may have far-reaching effects for farmers – as regards their access to seed and propagating material and their role as custodians of crop genetic diversity.

Under UPOV, plant breeders' rights shall be granted when a variety is new, distinct, uniform and stable. UPOV has developed minimum standards for plant breeders' rights, which member countries are required to implement. These standards have become stricter over the years. Since 1998, countries joining UPOV have had to accede based on the 1991 Act of the Convention, which is the strictest to date. Under the 1991 UPOV Convention, farmers are not allowed to exchange and sell seed of protected varieties among themselves: indeed, they are basically not allowed to save and use farm-saved seed of protected varieties for the next harvest. However, governments may exempt small-scale farmers from the latter restriction provided that the farmers use such farm-saved seed only on their own land holdings. Further, this optional exemption must be implemented within 'reasonable limits' and in such a way as to safeguard the interests of the breeders. It may involve the payment of a licence fee to the breeder for the use of farm-saved seed, which is practised in many OECD countries but is less common in developing countries.

If this type of plant breeders' rights legislation is introduced in a country like Nepal, farmers using protected varieties will no longer be able to be active on the informal seed market. As a result, the number of farmers who continue to exchange and sell seed will decline – thereby also reducing the crop genetic diversity available to farmers.

Farmers usually use the varieties they consider best adapted to their own land holdings. They also mix varieties in order to develop new ones and thus may wish to use protected varieties. UPOV '91 may act as a barrier to this practice since it prohibits the development of varieties that are classified as essentially derived from protected varieties (Article 15.5).

The novelty criterion in particular may affect Farmers' Rights. Whereas a plant variety had to be new under the 1978 Act of UPOV, it is sufficient under the UPOV '91 Act to discover and develop a plant variety (Article 1). This means that a plant breeder can discover and further develop a farmers' variety and be granted plant variety protection on the result. Such plant breeders' rights may affect the use of the original variety as well. In order to ensure that farmers' varieties are not misappropriated in this way someone must file an opposition, and the burden of proof in such cases rests with the one who files the opposition. Importantly, if farmers are to be active in such cases, they must first be aware of the plant variety protection applications *and* be in a position to file opposition. For the many subsistence farmers of Nepal this is often not the case.

Earlier Acts of the UPOV Convention were less strict and provided a better balance between breeders' and Farmers' Rights. In that way, they safeguarded compensation and incentives for the important work of plant breeders while ensuring that farmers could continue their traditional practices of seed saving, use, development and exchange so important for the global pool of genetic resources.

The National Alliance for Food Security (NAFOS)

NAFOS was established in 1999. By that time, it had become easier for NGOs to operate in Nepal, after multi-party democracy had been introduced in 1991.[6] The government encouraged both international and national NGOs, and as a result the number of projects aimed at improving economic development and social welfare increased. However, such development projects tended to focus on providing services, rather than reviewing the policies and legal frameworks being introduced by the government. This became evident when, after the establishment of NAFOS, one of the first things the member organisations did was to hold a series of meetings where they mapped the entire collection of donor-driven projects in Nepal within the area of food security and natural resources management to get an overview of what was being done by whom. This helped them identify gaps and needs as well as options for addressing these. Through this process they discovered that what was most lacking was policy advocacy, as the NGOs generally focused on service delivery. Further meetings were then held where they tried to specify the exact advocacy needs.

One particular concern, according to the NAFOS initiators, was that the issue of the right to food was not being dealt with, and Nepal continued to experience severe food security problems. Additionally, in 1998 the government had decided to apply for WTO membership. This had worried some NGOs who feared that the membership terms to be negotiated would impact negatively on national food security. They felt that there was a need to help the negotiators finalise a good accession package. Concluding that collaborative efforts were needed to strengthen Nepal's position during WTO membership negotiations, the NAFOS initiators decided to join forces and undertake a series of research, sensitisation, capacity-building and advocacy activities in rural and urban communities.

The most central organisations when NAFOS was created were ActionAid Nepal, the Forum for Protection of Public Interest (PRO PUBLIC) and the South Asia Watch on Trade, Economics and Environment (SAWTEE). It was ActionAid Nepal and SAWTEE that brought up the idea of establishing a network, and these two organisations have remained important to NAFOS. In addition, SAWTEE has been functioning as the NAFOS secretariat since 2001.

After NAFOS had been established, its various member organisations started implementing activities and issuing publications under the banner of the network, which allowed the network members to build up name recognition and capitalise on their common strength. Several members have mentioned how the cooperation in NAFOS made them stronger.

NAFOS advocacy strategy

The overall objective of NAFOS is to create more awareness about the right to food, Farmers' Rights, trade, biodiversity management and livelihood issues, with regard to stakeholders at the grassroots level as well as policy-makers. In this way the network hopes to contribute to increased food security in Nepal. The alliance believes this will be possible if they can succeed in creating a critical mass, at the grassroots and among decision-makers, that supports the right to food and the rights of farmers to a livelihood in Nepal in the context of globalisation, liberalisation and the WTO. Through its member organisations, NAFOS works at the grassroots level and at the policy level, to ensure that the concerns of the grassroots level are addressed in policy-making and implementation.

The network is held together by the common goal shared by all members: to achieve food security in Nepal. In addition, there are personal ties and friendship that make coordination easier. According to central individuals in the network, unity among the members and their ability to work together rather than promoting their own agendas constitute an important part of their identity as a network.

As a loose and action-oriented network, NAFOS can mobilise quickly when necessary but is usually more or less dormant. It does not have projects of its own and is not formally registered, but its key members actively conduct research on food security issues and share the results with each other for the purpose of potential advocacy programmes. Members regard the network as a platform for action and advocacy and as a forum where they can share their views and results. The 'No to UPOV' campaign marked a high point of activity for the network, and it has not been as active since. After the 'No to UPOV' campaign, NAFOS members have stayed in touch and key members meet quite frequently to discuss food security issues and the role of the network. Main activities of the alliance are policy and action research, sensitisation and capacity building, networking and alliance building and advocacy. Activities are jointly funded by the members.

Previously, the government had approached NGOs on an individual basis. In general, the various government branches were not overly eager to involve the NGO community in policy work, as they regarded NGOs as negative and not capable of coming up with constructive solutions for the challenges at hand. After the establishment of NAFOS, its member organisations always notified the rest of the network if they were approached by the government regarding an issue. NAFOS would then discuss the matter, make a joint assessment and reply as a network. According to central NAFOS members, the fact that NAFOS provided actual input to the policy process rather than merely criticism, gave them credibility. Because of their own capacity constraints and the growing realisation that NGOs can be development partners, the government came to need and appreciate such input. This way NAFOS did advocacy as a collaborative effort. Because its various member organisations had different institutional relations with the various ministries as well as relationships with individual officials, the network had considerable access to decision-makers. When NAFOS was established there were several other NGO networks in Nepal as well, but none within the same subject area. As the business actors were already well organised, it was seen as important to create a counter-weight.

As is natural for a diverse group of organisations, NAFOS members differ in their views on many issues. For this reason there has been talk about creating a 'task force' within NAFOS, made up of like-minded organisations. This has not happened yet, however, and for the 'No to UPOV' campaign it was unnecessary, as the members were all in agreement on that issue. NAFOS has so far worked only on issues all the members agree on, and the members do not want the network to be used in a way that creates division among the various organisations.

The network has also debated changing its structure. Some members would prefer NAFOS to become a more established institution, perhaps some sort of formally registered and action-oriented organisation. But as there are differing views on this within the network, NAFOS has so far remained loosely organised. Those in favour of maintaining the status quo argue that loose networks have some important advantages such as their inherent flexibility. When NAFOS was established as a loose network in 1999, one of the reasons this structure was chosen was because it was an easy way of organising the collaboration.

Originally the secretariat was meant to shift on a yearly basis, with each organisation serving as the secretariat for one year, to provide all the members with a feeling of ownership. It was also thought that this solution could provide the network with long-term viability thanks to the high number of members. However, this solution never materialised as most of the member organisations have been reluctant to take on this role and are happy to let SAWTEE continue in this role. Most members see it as important that the focal point at the organisation hosting the secretariat has the necessary skills and capacity to coordinate the network and drive the work forward and also that the organisation in question is able to provide the space needed for NAFOS. Not least because of its size and capacity, SAWTEE is regarded by the other members as well-suited to handle the responsibility of housing the network secretariat, and this solution has worked quite well. However, in the long run it might be possible to create more of an ownership feeling among the other members if they take turns serving as the secretariat.

Organisations behind NAFOS

NAFOS membership is open both to organisations and individuals working with natural resources management, trade and food security, and the members agree upon the importance of linking the realities on the ground with policies. Currently, NAFOS is made up of more than 20 local, national and international organisations.

One of the initiators and key members of NAFOS, SAWTEE, is itself a network. It was launched in 1994 in Nepal by a consortium of South Asian NGOs, and its secretariat is located in Kathmandu. Altogether this regional network has 11 member organisations from Bangladesh, India, Nepal, Pakistan and Sri Lanka. Its overall objective is to build the capacity of concerned stakeholders in the region within the areas of liberalisation and globalisation. The network wants to enable South Asian communities to benefit from changing regional and global economic and environmental trends, as well as to minimise the harm that these may cause.[7]

The other initiating organisation, ActionAid Nepal, already had a good collaborative relationship with SAWTEE when NAFOS was created, and the two organisations see themselves as partners. ActionAid Nepal is a national branch of ActionAid International, an international anti-poverty agency whose aim is to fight poverty worldwide. ActionAid has been working in Nepal since 1982; in 1998 the organisation adopted a rights-based approach aimed at creating an environment in which the poor and excluded can exercise their rights as well as deal with and overcome the causes and effects of poverty. Thus, the right to food became a top priority for the organisation.[8]

Other members of the NAFOS network include Caritas Nepal, Pro Public, Green Energy Mission Nepal, LI-BIRD, Nepal Permaculture Group, USC Canada and Care Nepal. As part of their collaboration within NAFOS, the member organisations have shared their experiences and created a link between grassroots initiatives and advocacy work. This division of labour, with some member organisations specialising in policy, while others mainly focus on livelihood enhancement at the community level, has allowed the network to make the most of the different experiences and areas of expertise of its members.

The WTO accession process and the 'No to UPOV' campaign

When the WTO came into being in January 1995, Nepal had been an observer country of the General Agreement on Tariffs and Trade (GATT) since 1993. Nepal therefore received observer status in the new organisation and also decided to apply for membership. Unlike the mostly automatic membership process of many other international organisations, accession to the WTO involves a process of negotiation. After the applicant government has submitted a formal written request for accession, the WTO General Council establishes a Working Party to examine the request and submit their findings to the General Council for approval. All members of the WTO can choose to join the Working Party.

The Working Party initially established in connection with Nepal's application for GATT membership in 1989, was continued after January 1996 to facilitate the country's WTO accession process. In 1998, Nepal submitted the required Memorandum on Foreign Trade Regime covering all aspects of the country's trade and legal regime, including economic policies, trade regulations and intellectual property rights (IPR) policies. This was then examined by the Working Party, and as part of the fact-finding process Nepal received altogether 364 questions and clarifications regarding the memorandum in January 1999. It responded to these in June 1999, but after the first formal meeting of the Working Party in May 2000, Nepal had to respond to additional questions raised by WTO members.

After all aspects of the trade and legal regimes then existing in Nepal had been examined in this way, the terms and conditions of WTO membership were negotiated at the multilateral level. Parallel to these negotiations, however, Nepal was also engaged in bilateral negotiations with interested Working Party members. It was as part of these bilateral negotiations that the issue of UPOV membership was brought up. It is quite common for what are often called 'WTO-plus' conditions to be introduced during the bilateral negotiations and pressure may be exerted to make the acceding country accept them.

On 9 August 2003, when Nepal was at the final stage of accession negotiations, the Nepalese government received a letter from the WTO Secretariat indicating that the country would have to join UPOV to fulfil its obligation to provide plant variety protection under the Agreement on Trade-Related Aspects of Intellectual

Property Rights (TRIPS). This caught the Nepalese government by surprise, as the issue had not been mentioned previously. Indeed, up until that point the delegation had felt that the process of WTO accession was moving along rather smoothly. The government informed SAWTEE about the letter the same day, seeking the organisation's input regarding the costs and benefits of UPOV membership for Nepal as the government delegation had to leave for Geneva the next day (10 August) to finalise the accession deal. They asked SAWTEE for a report on this question, which SAWTEE prepared and submitted to the government within a few hours. The conclusion of this brief report was quite clear: Nepal should *not* join UPOV. Doing so would reduce its policy space to protect Farmers' Rights, promote local agriculture and farming systems, and work towards food security. Following SAWTEE's report, the authorities expressed their commitment to reject UPOV membership. The next day, the government delegation, consisting of around a dozen officials and experts from the ministries, went to Geneva to finalise the accession deal. Then, on 11 August, SAWTEE called a NAFOS meeting. It was at this meeting that the network came up with the 'Say No to UPOV' campaign strategy, deciding to hold a press conference, to publish and circulate posters, to write articles for the national newspapers and to follow up the trade negotiators so as to counteract any pressure to join UPOV.

In Geneva, the Nepalese delegation was headed by the then Ministry of Industry, Commerce and Supplies (now the Ministry of Commerce and Supplies). Once the delegation arrived in Geneva for the last round of accession negotiations, they were taken directly from the airport to the meeting, without being presented with the opportunity to rest after the long flight, and the first session of the negotiations continued until late in the evening. As a result, the members of the Nepalese delegation were very tired, while their counterparts were relaxed and prepared.

The Nepalese response to the suggestion that Nepal should join UPOV was that they would have to look into the matter to see if this was in the country's interest or not. They were then asked who had told them to say that, and that they would have to join UPOV in order to comply with their TRIPS obligations. UPOV was presented as the sole viable *sui generis* option and because the 1991 version is the only version available to new members, this agreement was the one they needed to join. The Nepalese delegation asked why some UPOV members had not joined UPOV 1991 if this was the best option and tried to win time by saying that they did not know enough about the issue to make a decision. They claimed that they needed more time to consider the matter and did not say either yes or no.

While the Nepalese delegation was negotiating in Geneva, NAFOS tried to create attention to the story at home – and succeeded so well that on 14 August, the UPOV issue was covered by all the national newspapers. Central to this was the press conference organised by NAFOS on 13 August. Three speakers from three different member organisations spoke at this press conference: Prem Dangal from All Nepal Peasants' Federation, Ratnakar Adhikari of SAWTEE and Yamuna Ghale of ActionAid Nepal. One important goal of the press conference was to send the message that the concerned stakeholder groups in Nepal, including farmers, were against Nepalese UPOV membership. Accompanying the press conference was a

Figure 14.1 Newspaper headlines from 14 August 2003
    Source: SAWTEE

press release that NAFOS had issued to explain why they opposed UPOV membership for Nepal. The goal was to strengthen the government and their position in the negotiations, rather than to put pressure on them – after all, the government had already said that they did not want Nepal to join UPOV.

When the delegation received news of the campaign this gave them the opportunity to say no to joining UPOV by referring to national opinion. They presented the newspaper headlines as evidence that it would not be feasible for the Nepalese government to agree to join UPOV, explaining that under such circum-stances they could not return to Nepal with an agreement that included UPOV membership. In this way, the campaign provided the delegation with crucial and massive support from Kathmandu, which ultimately had a direct impact on the negotiations.

Involvement of farmers and other stakeholders

Although the campaign itself was organised very quickly and only lasted a few days, NAFOS and its member organisations had also worked on the issue of UPOV prior to the campaign through holding farmers' workshops in various areas (among other things). USC Nepal, for example, had organised workshops with farmers' groups and community organisations to talk about UPOV and the upcoming WTO access-ion negotiations. These workshops had showed that the farmers were basically unified in their view: they wanted the Nepalese delegation to promote and protect the rights of farmers to save and exchange seed during the negotiations. Caritas Nepal also brought work on UPOV to the district and community level, and the views of farmers were transmitted to the network. Such activities provided the alliance with legitimacy and a greater community basis and allowed them to speak on behalf of the farmers to a greater extent. According to NAFOS members, the network would not have been able to launch the 'No to UPOV' campaign without this previous mobilisation and networking, which had provided them with the necessary farmer support. Moreover, the preparedness and coordination of the network enabled it to organise the campaign in just one week.

Prior to the actual campaign, NAFOS had also worked with the business sector and the media, and this provided them with support when UPOV came up as an issue during the WTO accession negotiations. Several journalists had conducted research on behalf of NAFOS members and in this way learned about various aspects of food security. When the 'No to UPOV' campaign was initiated, these journalists were therefore quite knowledgeable and sympathised with the NAFOS agenda. Many felt a high degree of ownership to the issue since they had discovered the importance of seed sharing for Nepalese farmers and food security themselves through their research. As a result, the 'No to UPOV' campaign was extensively covered in the media.

To strengthen their relationship with the business sector, NAFOS invited representatives from the Federation of Nepalese Chambers of Commerce and Industry to meetings. At these meetings the NAFOS positions were presented. Reactions were generally positive, and it was agreed that development concerns would have to be addressed in an IPR regime.

The documentation provided by NAFOS – reports, newsletters, assessments and analyses – was instrumental in informing all stakeholders, including the media and the business sector, about the process as well as the possible consequences of UPOV membership for Nepal.

Nepal decides not to join UPOV

On 15 August 2003, the Nepalese delegation declined to join UPOV. In fact, their rejection was couched in very diplomatic terms: the country would consider becoming a member while taking into account national needs and priorities. This happened after discussions within the delegation about UPOV, and there was general agreement that UPOV membership would be disadvantageous to Nepal.

The next day the accession package was finalised. The terms of accession were later approved by the Fifth WTO Ministerial Conference in Cancun, Mexico, in September 2003 and Nepal was offered membership. In 2004, the country was there-fore able to formally join as the 147th member of the World Trade Organization.

Among the countries acceding to the WTO at that time, Nepal was the only one to reject UPOV membership. China, Vietnam and Cambodia had all agreed to join. Most of the Nepali stakeholders feel that the government's position with regard to UPOV as well as its ability to stand by it during the negotiations had a lot to do with the 'No to UPOV' campaign organised by NAFOS.

Nepal has until 1 July 2013 to implement the TRIPS Agreement.[9] This means that plant variety protection legislation is not required until then – a point that has provided the country with much time to consult and enter into dialogue with concerned stakeholders. Because the authorities are seeking their input and listening to their advice as far as possible, the organisations have not felt the need for a new campaign on this issue. As part of the drafting process, NAFOS members like SAWTEE and LI-BIRD are also disseminating information among local communities.

Why was the campaign a success?

The 'No to UPOV' campaign in Nepal was successful for various reasons, and the story offers many useful lessons. One important factor was the legitimacy enjoyed by the campaign and its organisers. Because some of its member organisations focus mainly on service delivery and community work, while others engage mostly in advocacy and policy work, NAFOS was able to draw upon a wide range of expertise and practical experience for the campaign. This division of labour meant that the network could link these two important aspects of food security efforts. NAFOS members both disseminated information to farmers and brought their concerns further. These farmer consultations served to raise the legitimacy of the network because they allowed NAFOS to speak with authority on behalf of the farmers of Nepal. The legitimacy they gained from this increased their impact.

Another key factor that contributed to the success of the campaign was NAFOS's characteristics and preparedness. Because the network was highly coordinated and had already worked on trade and IPR issues in relation to food security, it could mobilise quickly and forcefully when the occasion demanded it. Its loose organisational structure might have benefited the network in this situation. As seen by many of its members, one important feature of the network is the tolerance, unity and solidarity among the member organisations. From the time NAFOS was first initiated, trust building was seen as essential. The fact that the members stood together rather than promoting their individual agendas, that they shared information and communicated jointly with the government, made the network and their message stronger. It was also important that credit was given to the alliance, which further bolstered trust within it. The members focused on what they agreed on and regarded the diversity of opinions and views as something positive.

An important lesson to be drawn from this is that cooperation between NGOs is crucial. By avoiding a situation where a competitive environment could take root among the various organisations and focusing instead on collaboration and the sharing of information, NAFOS showed that it was possible to present a united front on issues of shared concern.

Another important factor that allowed NAFOS to launch the campaign was their communication with government representatives. NAFOS member-organisations had connections with government officials in various ministries and departments, and this access to information meant that they were well-informed and could act when needed. Communication with the authorities went both ways, as NAFOS and the various network members would also supply the government with information, analysis and recommendations. The government's responsiveness and willingness to listen was an important part of this. Interacting with the authorities in this way was important in terms of creating relations based on trust and demonstrating that the NGO community could engage, be proactive and constructive.

This trust proved central: without the trust of the government, NAFOS would not have received crucial information about the WTO negotiations. Understandably, the government felt that it had to balance the need to inform and cooperate with civil society with the need for secrecy. For the purpose of negotiating strategies, they could not share the negotiation mandate or disclose how far they might be willing to go. However, because they trusted NAFOS members like SAWTEE and ActionAid to keep the necessary details confidential, they shared some information with these organisations and through them the rest of the NAFOS network. This enabled NAFOS to act. One lesson from this is that in order to create a trust-based relationship with the government, NGOs might have to demonstrate their willingness and ability to interact constructively and such a relationship can prove instrumental to NGO impact.

The media strategy was also a very important part of the campaign and another reason for its success. NAFOS had a deliberate and well thought out media strategy, and because the network succeeded in getting the story out and the campaign received such a substantial amount of attention, it was possible for the Nepalese delegation to use domestic opinion regarding UPOV membership as an explanation for not committing to join. Since NAFOS was a large alliance, it was probably also easier for them to be heard than it would have been for individual organisations. NAFOS's strategy in this respect teaches us that an effective media campaign can be a valuable means for awareness-raising and mobilising the public, increasing the likelihood of impact.

Interaction with the media and the government formed two parts of NAFOS's three-level strategy: they were also communicating with various actors from the private sector, such as other NGOs and the business sector. This strategy, the involvement of various parts of the Nepalese society and seeking the support of central stakeholders was seen by many as central to the campaign's success. An important lesson from this is that involving as many stakeholders as possible, and engaging in consensus-building efforts, can increase the likelihood of success.

It is also likely that NAFOS had more of an impact than any of the individual organisations would have had by themselves because the joint views of a relatively large and diverse alliance holds more sway and enjoys greater legitimacy than in the case of a single organisation on its own. Another factor that probably increased the impact of the network was their expertise on the subject matter. Research had been carried out by both NAFOS members and the network itself, as well as reviews of policies and plans. Engaging university researchers was an important part of this work, and in this way the contact network of NAFOS was expanded further. In general there was a high level of trust in their expertise and that they wanted what was best for Nepal.

NAFOS members also cite the heartfelt conviction of central individuals as important to the success of the campaign. The passion and conviction of the campaign organisers may well have had a role in the successful outcome. Some of these organisers had experience from abroad or worked for organisations based outside Nepal, so an additional explanation for the network and the campaign strategy might be that they received inspiration from abroad. One lesson from this is that individuals can have considerable impact on organisations, networks and their actions; further, that in situations where organisations want to take action they may have to come together to effect change.

It can be argued that the 'No to UPOV' campaign contributed to the democratic development in Nepal, as one of the results was greater confidence in the government because the farmers felt that they were listened to and that their concerns were taken into account by the authorities. This shows that campaigns with broad participation from various sectors can have substantial impact, also beyond the case in focus at the time.

The campaign also led to better understanding and awareness of the importance of the right to save and exchange seeds and the links between seed practices, biodiversity and food security. In addition, the campaign was central to the integration of these issues in the action plans and project implementation of NAFOS's member organisations, and it this way benefited other aspects of the maintenance of crop diversity.

Concluding remarks

Although it is unlikely that the precise circumstances of this story will be replicated in another context, it provides some valuable lessons on how organisations involved in the maintenance of crop genetic resources and the promotion of Farmers' Rights can influence official policies through networking, collaboration and communication, within civil society and with the government and the media – and how they can give farmers a voice.

In societies with numerous NGOs, coordination of efforts and a feeling of solidarity can be crucial if common goals are to be realised. If there is distrust between the government and civil society and no systems for information and feedback regarding laws and policies have been institutionalised, establishing a trust-based relationship with the government becomes essential for NGOs that seek to have an impact on policy and legislation.

This story also shows that the rights of farmers to save, use and exchange seed can be protected and promoted also in countries that have to fulfil their TRIPS obligations and that it is possible to accede to the WTO without committing to join UPOV. This is a point worth bearing in mind especially for those developing countries wishing to become WTO members but who are worried about the consequences for food security upon joining UPOV.

Notes

1   Under the TRIPS Agreement, least-developed countries were given a transition period for the introduction of protection for trademarks, copyright, patents and other intellectual property. Originally, this transition period was to expire on 1 January 2006. However, following a decision reached by member governments on 29 November 2005, least-developed countries were given an extension until 1 July 2013 to provide such protection (see www.wto.org/english/news_e/pres05_e/pr424_e.htm). This means that Nepal has until 1 July 2013 to introduce plant variety protection. So far, the country has not introduced such legislation.

2   This chapter is based on information from interviews with representatives of NAFOS member organisations and other central stakeholders conducted in Kathmandu, Nepal in February 2009.

3   The 2011 Human Development Index ranks Nepal as the 157th of 187 countries and therefore in the 'low human development' category. Available from www.hdr.undp.org/en/media/HDR_2011_EN_Table1.pdf

4   According to the CIA World Fact Book (www.cia.gov/library/publications/the-world-factbook/geos/np.html) the urban population comprised only 19 per cent of the total population of Nepal in 2010, but the annual urbanisation rate is believed to be 4.7 per cent until 2015.

5   Public plant breeding is a big sector in India.

6   A multi-party parliament took seat in May 1991 after constitutional reforms.

7   See the SAWTEE website for more information www.sawtee.org/

8   See the ActionAid website for more information www.actionaid.org/

9   Following the WTO decision of 29 November 2005, to extend the transition period for least-developed countries, Nepal has until 1 July 2013, to introduce plant variety protection (see Note 1).

Part VI
Ways forward

15 Future prospects for the realisation of Farmers' Rights

Regine Andersen and Tone Winge

This book has shown that some success has been achieved with regard to all the elements necessary for the realisation of Farmers' Rights as they are recognized in the Plant Treaty. We have seen how it can be possible to uphold or create legal space for farmers to save, use, exchange and sell farm-saved seed even when national commitments to regional or international agreements provide detrimental conditions. Moreover, it is possible to promote the sharing of traditional knowledge related to crop genetic resources among farmers while also avoiding the possibilities for misappropriation. There are many encouraging examples of benefit sharing, with potential for being scaled up for greater outreach and more beneficial impact for farming communities. And we have seen promising examples of farmers' participation in decision-making at the national level, with important lessons to be derived. The success stories presented in this book give an idea of the wide range of prospects for realising Farmers' Rights and the positive effects for the conservation and sustainable use of crop genetic resources and on the livelihoods of farmers around the world. Not least, they show that Farmers' Rights are already being implemented.

In this concluding chapter we focus on some central factors of success, indicating some gaps and needs, and discuss the future prospects for the realisation of Farmers' Rights, at the national and international levels.

Some central factors of success

The factors of success for the stories presented in this book differ precisely because the stories and their framework conditions vary so greatly. That is why we decided to highlight the most important factors in each story. On the other hand, we can also note some factors that are common to many of these stories.

Perhaps the most important factor of all is taking farmers' perceptions and realities as the point of departure for action. Only when the Food Safety Authority of Norway understood the concerns of farmers who were engaged in crop genetic diversity related to variety release and seed marketing regulations were they able to the amend regulations so as to accommodate the most important needs of these farmers. Only when farmers are actively engaged in participatory plant breeding in their own fields and with their own needs and priorities as points of departure, will plant breeding truly meet their needs and build on their knowledge – as we saw, for example, in the case from Syria. Only when farmers' views are taken as the basic starting point will advocacy campaigns on Farmers' Rights have legitimacy. Supporting farmers' own initiatives or finding ways and means to involve farmers in the planning and implementation of action related to Farmers' Rights thus emerges as a central factor of success.

The work done by small organisations, like the Basque Seed Network, shows that even relatively small-scale initiatives and projects can contribute to the maintenance of crop genetic diversity and the realisation of Farmers' Rights, providing examples and models for others.

An essential lesson is that activities to promote the conservation and sustainable use of crop genetic resources among farmers will succeed only if such activities provide prospects of improved livelihoods – as shown not least by the case from Nepal. By developing a market for traditional rice varieties, farmers in the areas around Pokhara achieved great improvements in their livelihoods. The conservation and sustainable use of crop genetic resources became a foundation for prosperity. The conservation and sustainable use of crop genetic resources must be shown to be economically viable for farmers. This can be achieved through access to markets, as in Nepal, or through incentive structures and support mechanisms that compensate for farmers' contribution to the genetic pool, as being considered in Norway. And it can also be achieved simply by providing food that is healthier, tastier and more filling, as we saw in the Philippines.

We have also seen that involving breeders and scientists can greatly enhance the prospects of success, especially if the projects take as their points of departure the actual situation in the field and farmer priorities. This points to a highly promising potential of benefit sharing which could be developed to a much greater extent and could contribute to enhancing farmers' livelihoods by use of crop genetic diversity.

Another recurring factor has been the role of civil society organisations (CSOs). In many of the stories presented here, CSOs have had a central role as facilitators and/or initiators – perhaps because CSOs often work closely and directly with farmers and tend to know their needs. CSOs have been important contributors to the realisation of Farmers' Rights. In addition to functioning as initiators of projects and facilitators of collaboration among other actors, various CSOs have played a vital role in advocacy work, awareness-raising, disseminating information and building capacity.

Local ownership is vital, also in institutional terms. The importance of involving local institutions has also been underlined by Jarvis and Hodgkin (2008). Such local involvement can be crucial in promoting sustainability and continuity as well as in efforts to scale up activities. As the story from Syria shows, government attitudes and willingness to cooperate can be central, especially with regard to involving the appropriate national institutions. The Philippine case also demonstrates the importance of local ownership through involving the local authorities in order to ensure sustained success.

And finally we note the importance of the funding source of benefit sharing. In addition to the Benefit Sharing Fund of the Plant Treaty, development cooperation can be an important source of benefit sharing. In Chapter 1 we saw how the realisation of Farmers' Rights is crucial to poverty alleviation and rural development. Thus, donor agencies that strive to fight poverty in rural areas would do well to contribute to the realisation of Farmers' Rights. In a very literal sense, plant genetic diversity keeps the poor alive – so it makes sense to invest in improving that potential. To scale up good local models to the national level requires capacity building and funding. Implementing Farmers' Rights requires support, and here we may note that the Plant Treaty foresees development cooperation as playing an important role.

Current gaps and needs

In order to discuss the future prospects for the realisation of Farmers' Rights, we first need to identify the main gaps and needs. These are summarised, based on the work of international informal surveys and consultations referred to in Chapter 1 and the deliberations in this book.

Gaps and needs: farm-saved seed

An urgent challenge is to protect and enhance farmers' legal space to save, use, exchange and sell farm-saved seed. This was an issue of controversy throughout the years of negotiations leading to the Plant Treaty and also in the subsequent sessions of the Governing Body. Now a shared understanding of the importance of legal space has emerged among all central stakeholder groups, as a result of greater awareness of the vital contributions of farmers to the conservation and sustainable use of crop genetic resources and increased attention to the detrimental effects of certain forms of legislation. However, opinion differs as to what legal space for farmers should cover in this context.

Regulations on variety release and certification of seed and plant propagating material pose serious hurdles to the rights of farmers to exchange and sell farm-saved seed in many countries. Such regulations often obstruct farmers in continuing on-farm conservation and the sustainable use of crop genetic diversity. Some of the arguments for these seed policies have centred on the need to ensure plant health and seed quality – but this argument is short-sighted. In the longer run, plant health and seed quality will depend on the availability of diverse plant genetic resources. Landraces and other traditionally used varieties can provide humanity with precisely such vital diversity. Such varieties often dispose of a broad and diverse genetic base, which enable them to better adapt to environmental challenges than genetically homogenous varieties (Robinson, 1996). We need to develop shared norms on how seed laws can be designed so as to ensure adequate legal space for farmers. One possibility could be to allow an informal market for seed from unregistered varieties to exist alongside the formal market.

Intellectual property rights also constitute barriers to the realisation of Farmers' Rights, to varying degrees. In some countries, the balance between farmers' and breeders' rights is seen as acceptable, as in India and in Norway today. Elsewhere, plant breeders' rights and patents constitute greater hurdles, as customary uses and exchange of seed are prohibited. We need to examine and discuss the kind of legal space that farmers need with regard to plant breeders' rights and patents, with a view to developing shared norms.

Gaps and needs: protection of traditional knowledge

In several countries, fear of misappropriation of farmers' varieties and associated traditional knowledge has led to farmers becoming protective of their knowledge and seeds. This development is detrimental to farmers' sharing of seed and knowledge as well as to *ex situ* conservation measures. We will need to find ways and means of ensuring that farmers do not need to fear misappropriation. One challenge is to identify efficient measures for establishing *prior art* for landraces and farmers' varieties in order to ensure that these cannot be made subject to intellectual property rights. Our cases from the Philippines and Peru have shown how this may be done. A further challenge is to get provisions included in laws on intellectual property rights to ensure that no misappropriation takes place.

Just as both *in situ* conservation and *ex situ* conservation are necessary to ensure that plant genetic resources for food and agriculture are maintained for the benefit of future generations, both the documentation and practice of traditional knowledge are needed, if what remains of this knowledge is to survive. Ageing farmer populations and rules stipulating that only seed from registered varieties that fulfil certain criteria can be marketed are among the factors that threaten the practice of seed saving/exchange and the related traditional knowledge. As we have seen in several chapters, the challenge of ageing of the farmers and gardeners who still work to maintain the plant genetic diversity is a widespread concern: for many reasons, young people are abandoning the rural areas or lack interest in the maintenance of crop genetic diversity. How then can agriculture and crop genetic diversity be made more appealing to the young generation?

Gaps and needs: rights to participate in benefit sharing

There are many good examples of the realisation of Farmers' Rights at the local level, such as benefit-sharing measures and measures to protect traditional knowledge from extinction. Many of these have the potential to be scaled up to the regional or national level; for example, through extension service systems – but there have as yet been no examples. Further consideration is required on how to facilitate and support such expansion.

A central challenge in the coming years will be to scale up successful models and initiatives while also adapting them to local circumstances in collaboration with local institutions and farmers. This will require a certain amount of human and financial resource but also laws and policies that do not create barriers to central activities like the distribution of seed. In many countries the restrictions put in place by seed legislation pose a barrier to the realisation of Farmers' Rights. Confusion about the provisions of such legislation may also create difficulties for projects that rely on the distribution of seed. For example, for projects aimed at developing new varieties in collaboration with farmers, it is necessary that the seed laws allow the necessary amount of seed to be multiplied and distributed.

Moreover, major incentive structures and regulations in agriculture are often detrimental to the on-farm conservation, sustainable use and further development of crop genetic diversity and represent serious hurdles to the full realisation of Farmers' Rights. What is promoted is often the use of high-yielding genetically homogeneous crops, with loans, milling, processing, marketing and other business services available to those who grow high-yielding varieties – but not to small-scale farmers who still grow traditional, genetically highly diverse varieties of crops. Efforts to amend these incentive structures to also accommodate the needs of farmers engaged in crop genetic diversity are urgently needed. For example, extra support could be provided per area cultivated with certain traditional or modern farmers' varieties, and there could be rewards for farmers who make particular contributions to the global genetic pool.

Gaps and needs: participation in decision-making

Farmer participation in decision-making is a many-faceted issue. In general, in countries where farmers are granted some sort of participation, those engaged in the conservation and sustainable use of crop genetic diversity are rarely represented. Ways and means to identify such farmers and get them involved in decision-making processes are needed: awareness-raising and capacity building are important measures in this regard.

Future prospects for the realisation of Farmers' Rights at the national level

How can Farmers' Rights be implemented at the national level according to the provisions in the Plant Treaty? How can an implementation process be designed that enables countries to take the measures they deem necessary, in line with their own needs and priorities? Some steps have been taken to address various elements of Farmers' Rights, but only in very few countries (among them the Philippines, Peru and Norway) have processes begun to implement Farmers' Rights more systematically. There is thus only limited experience to draw on. It is important to seek to identify the possible avenues to a more systemic approach to realising Farmers' Rights in line with countries' own needs and priorities. Awareness-raising and consultation processes to identify such avenues are important measures here.

Creating and enhancing awareness on the importance of Farmers' Rights

In most countries, awareness of Farmers' Rights and their importance for food security and poverty eradication is marginal. Thus, a point of departure for the implementation process must be to create or enhance awareness on the importance of these rights, on their contents and background. Seminars and workshops are useful instruments. Participants could be asked to define Farmers' Rights within the context of the specific conditions of their own country and to highlight the rights they deem most important. They could also be challenged to explain why these rights are so important. The division of labour among various institutions and sectors could also be covered at such seminars and workshops. Finally, participants could be spurred to see how they could join forces and pool resources towards the realisation of Farmers' Rights.

There are many ways to organise and conduct seminars and workshops. They may be held in central organisations such as farmers' organisations, parliamentary committees vested with agriculture, relevant departments/units and/or agencies in the ministry of agriculture, institutions involved in extension services to farmers, relevant research institutions, NGOs engaged in crop genetic resources and Farmers' Rights, and central seed corporations and/or plant breeding institutions. They may be organised sector-wise for all members of an organisation or institution. Seminars and workshops could also be organised for representatives from the central organisations and institutions in each of the sectors pertaining to Farmers' Rights.[1]

Creating awareness through the media is a way of reaching out to broader target groups – such as farmers (with access to television, radio or newspapers/magazines), consumers, various interest groups and stakeholders and the public at large. Getting the media interested in Farmers' Rights can be a challenge. Often it is useful to identify entry points in the hot topics of the day, like the food crisis, climate change or perhaps new acts of legislation up for debate and to use these entry points to make clear the crucial relevance of Farmers' Rights.

Ensuring farmers' participation in decision-making

There is as yet not much tradition for involvement of farmers in decision-making processes, in particular farmers working with crop genetic resources. Certain questions need to be answered:

- Who are the farmers that should be included in the processes?
- Should all farmers be included on an equal basis or only smallholder farmers or perhaps only farmers who actively participate in the conservation and sustainable use of crop genetic diversity?
- How can their representatives be identified?
- What are legitimate processes for identifying representatives?
- Who should identify such representatives?
- Are there any other legitimate processes?
- How can farmers' participation in the implementation process be ensured?

Farmers' awareness of these questions and capacity building are key factors towards their participation in decision-making at the national level.

Developing a national consultative process and pooling resources

Whereas the first two steps above are concerned with the foundations of the implementation process, this third step describes how a broad-based national consultative process can be designed for the implementation of Farmers' Rights.

A broad-based consultative process should ensure participation from all central stakeholder groups, from all regions of the country, from different ethnic groups (if relevant) and from men and women. Depending on the size of the country and the resources available, this process may start out with workshops in various regions or administrative units of the country. Representatives of the different regions/units could then be invited to national-level workshops to present and represent the results (see e.g. Scurrah *et al.*, 2008).

A first national-level workshop could aim at outlining the contents and structures of a framework of implementation of Farmers' Rights. A second national-level workshop could be structured to detail these recommendations and develop strategies for safeguarding their implementation; for example, taking the gaps and needs above as points of departure for discussions. Further workshops could be conducted to monitor progress in implementation and provide recommendations as to the further steps required. A key challenge will be to design such workshops as true dialogues involving the stakeholders – discussions regarding Farmers' Rights have previously been beset with controversies and have brought little progress.

Future prospects for the realisation of Farmers' Rights at the international level

Although Article 9 of the Plant Treaty leaves responsibility for implementing Farmers' Rights to the national governments, this does not mean that the Governing Body of the Plant Treaty has no role to play. Article 21 stipulates that the Governing Body shall promote compliance with *all provisions* of the Treaty. Procedures and mechanisms towards this end are to include monitoring and offering advice or assistance, including legal advice or legal assistance, when needed. Article 19.3 specifies that the Governing Body is to promote the full implementation of the Treaty, through a range of measures, including plans and programmes. And the Preamble highlights the need to promote the implementation of Farmers' Rights at the national and international levels.

In light of the results from several informal international consultations and surveys prior to the Governing Body sessions of the Plant Treaty, some recommendations appear particularly important:

(1) First of all, the Governing Body may wish to consider establishing an *ad hoc* working group to propose voluntary guidelines for the implementation of Article 9 (and related provisions of the Plant Treaty) to the Governing Body, taking the gaps and needs highlighted above into consideration. (2) It should also encourage the Contracting Parties to develop national plans for implementing Farmers' Rights and to submit regular reports on the implementation of Farmers' Rights. (3) The Governing Body should facilitate guidance and assistance to Contracting Parties seeking such guidance and assistance with regard to the implementation of Farmers' Rights. (4) Furthermore, the Governing Body should encourage documentation of, and research on, the implementation of Farmers' Rights at the national level, in a wide range of countries, to facilitate the sharing of experiences.

Finally, the Governing Body should increase its efforts to attract the funding required for the implementation of Farmers' Rights. In particular, consideration should be given to measures for strengthening the Funding Strategy and for mobilising development cooperation with reference to Articles 7 (on national commitments and international cooperation) and 8 (on technical assistance) of the Treaty.

Concluding remarks

The realisation of Farmers' Rights is already underway, as the success stories presented in this book have shown. Will this process gain speed? That depends on awareness of the challenges, on political priority and on international cooperation. We have offered some practical models and lessons, embedded in a conceptual framework of Farmers' Rights, in the hope of inspiring decision-makers and practitioners to take action and work actively for the realisation of Farmers' Rights. This indeed is a matter of urgency – for food security and poverty alleviation, for development and society, indeed for our very future on Earth.

Note

1   Such organisations and institutions might be farmers' organisations, associations, groups and/or networks; parliamentary committees vested with agriculture and intellectual property rights; central relevant departments/units of relevant ministries in the country; the focal point for the Plant Treaty; institutions involved in intellectual property rights (plant breeders' rights and patents), variety release and seed certification bodies; research and capacity building institutions, including extension services, gene banks, relevant research institutions, relevant capacity building institutions, training centres, etc. relevant NGOs, other peoples' organisations, seed industry and its associations.

Bibliography

Abay, F. (2007) *Diversity, adaptation and GxE interaction of barley (Hordeum vulgare L.) varieties in Northern Ethiopia*, PhD thesis, Norwegian University of Life Sciences (UMB).

Abay, F. and Bjørnstad, Å. (2009) 'Specific adaptation of barley varieties in different locations in Ethiopia', *Euphytica*, 167(2), pp. 181–95.

Abay, F., Waters-Bayer, A. and Bjørnstad, Å (2008) 'Farmers' seed management and innovation in varietal selection: implications for barley breeding in Tigray, northern Ethiopia', *Ambio*, 37, pp. 312–20.

Abay, F., de Boef, W. and Bjørnstad, Å. (2011) 'Network analysis of barley seed flows in Tigray, Ethiopia: supporting the design of strategies that contribute to on-farm management of plant genetic resources', *Plant Genetic Resources*, 9(4), pp. 495–505.

Aderajew, H. and Berg, T. (2006) 'Selectors and non-selectors: agricultural and socio-economic implications of on-farm seed selection in Ethiopia', *Plant Genetic Resources Newsletter*, 145, pp. 1–10.

Adhikari, K. (2008) *Intellectual Property Rights in Agriculture: Legal Mechanisms to Protect Farmers' Rights in Nepal*, Kathmandu: Forum for Protection of Public Interest (Pro Public) and South Asia Watch on Trade, Economics and Environment (SAWTEE).

Almekinders, C.J.M. and Elings, A. (2001) 'Collaboration of farmers and breeders: participatory crop improvement in perspective', *Euphytica*, 122(3), pp. 425–38.

Andersen, R. (2005a) *The History of Farmers' Rights. A Guide to Central Documents and Literature*, FNI Report 8/2005, Lysaker, Norway: Fridtjof Nansen Institute.

Andersen, R. (2005b) *The Farmers' Rights Project: Background Study 2: Results from an International Stakeholder Survey on Farmers' Rights*, FNI Report 9/2005, Lysaker, Norway: Fridtjof Nansen Institute.

Andersen, R. (2008) *Governing Agrobiodiversity: Plant Genetics and Developing Countries*, Aldershot: Ashgate.

Andersen, R. (2009) *Information paper on Farmers' Rights submitted by the Fridtjof Nansen Institute, Norway, based on the Farmers' Rights Project*, input paper submitted to the Secretariat of the Plant Treaty 19 May 2009, (IT/GB-3/09/Inf. 6 Add. 3).

Andersen, R. (2012) *Plant Genetic Diversity in Agriculture and Farmers' Rights in Norway*, FNI Report 17/2012, Lysaker, Norway: Fridtjof Nansen Institute.

Andersen, R. (2013) Farmers' Rights in times of change: illusion or reality? In: de Boef, Walter S. *et al.* (2013), Chapter 6.2.

Andersen, R and Berge, G. (2007) *Informal International Consultation on Farmers' Rights, 18–20 September 2007, Lusaka, Zambia*, Report M-0737 E, Norwegian Ministry of Agriculture and Food, Oslo.

Andersen, R. and T. Winge (2008) *The Farmers' Rights Project – Background Study 7: Success Stories from the Realization of Farmers' Rights Related to Plant Genetic Resources for Food and Agriculture*, FNI Report 4/2008, Lysaker, Norway: Fridtjof Nansen Institute.

Andersen, R. and T. Winge (2011) *The 2010 Global Consultations on Farmers' Rights: Results from an Email-based Survey*, FNI Report 2/2011, Lysaker, Norway: Fridtjof Nansen Institute.

Andersen, R. and T. Winge, with contributions from B. Batta Torheim (2011) *Global Consultations on Farmers' Rights in 2010*, FNI Report 1/2011, Lysaker, Norway: Fridtjof Nansen Institute.

Ashby, J.A. (2009) 'The impact of participatory plant breeding', in S. Ceccarelli, E.P. Guimaraes and E. Weltzien (eds) *Plant Breeding and Farmer Participation*, Rome: Food and Agriculture Organization of the United Nations (FAO).

Bala Ravi, S. (2004) *Manual on Farmers' Rights*, Chennai: M.S Swaminathan Research Foundation.

Bellon, M.R. (2006) 'Crop research to benefit poor farmers in marginal areas of the developing world: a review of technical challenges and tools', *CAB Reviews: Perspectives in Agriculture, Veterinary Science, Nutrition and Natural Resources 2006*, 1(70).

BIOANDES (ed.) (2008a) *Diversidad de papas en el distrito de Pitumarca*, Cusco, ETC Andes: COSUDE, CEPROSI, BIOANDES.

BIOANDES (ed.) (2008b) *Variedades de papas nativas y conocimientos campesinos: microcuenca Shitamalca, San Marcos, Cajamarca*, Cajamarca, ETC Andes: COSUDE, Centro Ideas, BIOANDES [online]. Available from www.etcandes.com.pe/bioandes2/herramient ascomunicacion/CATALOGO%20PAPAS%20cajamarca.pdf [accessed 3 December 2012].

CBDC-Bohol Project and SEARICE (2002) *Profile of Farmer-Breeders in Bohol*, Quezon City: SEARICE.

Ceccarelli, S. (2012a) *Plant Breeding with Farmers – a Technical Manual*, Aleppo, Syria: ICARDA.

Ceccarelli, S. (2012b) 'Landraces: importance and use in breeding and environmentally friendly agronomic systems', in Maxted N., Ehsan Dulloo M., Ford-Lloyd B.V., Frese L., Iriondo J. and Pinheiro de Carvalho M.A.A. (eds) *Agrobiodiversity Conservation: Securing the Diversity of Crop Wild Relatives and Landraces*, Wallingford, UK: CABI Publishing International, pp. 103–17.

Ceccarelli, S. and Grando, S. (2002) 'Plant breeding with farmers requires testing the assumptions of conventional plant breeding: Lessons from the ICARDA barley program', in D.A. Cleveland and D. Soleri (eds) *Farmers, Scientists and Plant Breeding: Integrating Knowledge and Practice*, Wallingford, UK: CABI Publishing International, pp. 297–332.

Ceccarelli, S. and Grando, S. (2007) 'Decentralized-participatory plant breeding: an example of demand driven research', *Euphytica* 155, pp. 349–60.

Ceccarelli, S., Grando, S., Tutwiler, R., Baha, J., Martini, A.M., Salahieh, H., Goodchild, A. and Michael, M. (2000) 'A methodological study on participatory barley breeding. I. Selection phase', *Euphytica* 111(2), pp. 91–104.

Ceccarelli, S., Grando, S., Singh, M., Michael, M., Shikho, A., Al Issa, M., Al Saleh, A., Kaleonjy, G., Al Ghanem, S.M., Al Hasan, A.L., Dalla, H., Basha, S. and Basha T. (2003) 'A methodological study on participatory barley breeding II. Response to selection.' *Euphytica* 133(2), pp. 185–200.

CIP (ed.) (2006) *Catálogo de Variedades de Papa Nativa de Huancavelica – Perú*, Lima: International Potato Center (CIP), Federación Departamental de Comunidades Campesinas de Huancavelica (FEDECCH) [online]. Available from www.cipotato.org/publications/pdf/003524.pdf [accessed 3 December 2012].

CIPR (2002) *Integrating Intellectual Property Rights and Development Policy*, London: Commission on Intellectual Property Rights.

Correa, C. (1998) *Implementing the TRIPS Agreement. General Context and Implications for Developing Countries*, Penang: Third World Network.

Cosio Cuentas, P. (2006) *Variabilidad de papas nativas en seis comunidades de Calca y Urubamba –* *Cusco*, Cusco: Asociación Arariwa.

de Boef, W.S., Subedi, A., Peroni, N., Thijssen, M.H. and O'Keeffe, E. (2013) *Community Biodiversity Management. Promoting Resilience and the Conservation of Plant Genetic Resources*, Abingdon: Routledge.

Esquinas-Alcázar, J. (2005) 'Protecting crop genetic diversity for food security: political, ethical and technical challenges', *Nature Reviews Genetics*, 6, pp. 946–53.

FAO (1987) 'Report by the Chairman of the Working Group on its Second Meeting', *Report of the Second Session of the Commission on Plant Genetic Resources*, CL 91/14, Appendix F.

FAO (1998) *State of the World's Plant Genetic Resources for Food and Agriculture*, Rome: Food and Agriculture Organization of the United Nations.

FAO (2010a) *The State of Food Insecurity in the World: Addressing Food Insecurity in Protracted Crises*, Rome: Food and Agriculture Organization of the United Nations.

Fujisaka, S., Williams, D. and Halewood, M. (2010) *The Impact of Climate Change on Countries' Interdependence on Genetic Resources for Food and Agriculture. An Executive Summary*, Rome, Italy: Bioversity International.

Galié, A. (2012) 'Equal access for women to seeds and food security in Syria', *Food Security Insights*, January 2012, 82, Brighton, UK: IDS Knowledge Services.

Gautam, J.C. (2008) *Country Report On The State of the Nepal's Plant Genetic Resources for Food and Agriculture*, Rome: Commission on Genetic Resources for Food and Agriculture.

Ghislain, M., Spooner, D.M., Rodríguez, F., Villamón, F., Núñez, J., Vásquez, C., Waugh, R. and Bonierbale, M. (2004) 'Selection of highly informative and user-friendly microsatellites (SSRs) for genotyping of cultivated potato', *Theoretical and Applied Genetics*, 108(5), pp. 881–90.

Gómez, R. (2000) *Guía para las Caracterizaciones Morfológicas Básicas en Colecciones de Papa Nativas*, Lima: Centro Internacional de la Papa (CIP).

Grando S., von Bothmer, R. and Ceccarelli, S. (2001) 'Genetic diversity of barley: Use of locally adapted germplasm to enhance yield and yield stability of barley in dry areas', in Cooper, H.D., Spillane, C., and Hodgkin, T. (eds) *Broadening the genetic base of crop production*, New York and Rome: CABI, FAO, IPGRI.

Gutiérrez, R. and Valencia, C. (2010) *Las Papas Nativas de Canchis*, Intermediate Technology Development Group (ITDG), FONTAGRO: Lima.

Gyawali, S., Sthapit, B., Bhandari, B., Shrestha, P., Joshi, B.K., Mudwari, A., Bajracharya, J. and Shrestha P. (2006) 'Participatory plant breeding (PPB): A strategy for on-farm conservation and the improvement of landraces', in B. Sthapit, D. Gauchan, A. Subedi and D. Jarvis (eds) *On-farm Management of Agricultural Biodiversity in Nepal: Lessons Learned*. Proceedings of National Symposium, 18–19 July 2006, Kathmandu.

Halewood, M., Noriega, I.L. and Louafi, S. (eds) (2013) *Crop Genetic Resources as a Global Commons. Challenges in International Law and Governance*, Abingdon: Earthscan.

Helfer, L. (2002) *Intellectual Property Rights in Plant Varieties: An Overview with Options for National Governments*, Rome: FAO Legal Papers Online, FAO.

Hiroshima Prefecture Agricultural Gene Bank (1995) *Exploration and Collection of Plant Genetic Resources in Hiroshima Prefecture*, A Report of the Farm-to Farm Search Campaign (in Japanese).

Huaman, Z. and Gómez, R. (1994) *Descriptores de la Papa para Caracterización Básica de Colecciones Nacionales*, Lima: Centro Internacional de la Papa (CIP).

IFPRI, International Food Policy Research Institute (2010) *Ensuring Food and Nutritional Security in Nepal: A Stocktaking Exercise*, Kathmandu: USAID.

Iriarte, V., Condori, B., Parapo, D. and Acuña, D. (2009) *Catálogo etnobotánico de papas nativas del altiplano norte de La Paz – Bolivia*, Cochabamba: Fundación PROINPA

Japan Organic Agriculture Association (2010) *Reports on Production, Trade and Utilization of Seeds and Seedlings for Organic Agriculture No.2: On Seed Production by Farmers* (in Japanese)

Jarvis, D.I. and Hodgkin, T. (2008) 'The maintenance of crop genetic resources on farm: Supporting the Convention on Biological Diversity's Programme of Work on Agricultural Biodiversity', *Biodiversity*, 9(1&2).

Karki, S. (2004) 'Wild about rice', *Nepali Times*, 182, 6 February [online]. Available from www.nepalitimes.com/issue/2004/02/06/Leisure/10691 [accessed 26 November 2012].

Merino, R., Carballo, J., Vargas, F., Ortiz, N., Vargas, P., Rodríguez, E., Ortiz, M., Torrez, V., Carballo, F. and Vargas, D. (2004) *Catálogo de las variedades locales de papa y oca en la zona de la Candelaria*, Cochabamba and Lima: Fundación PROINPA and Centro Internacional de la Papa (CIP).

MINAG (ed.) (2008) *Papas nativas del Perú: catalogo de variedades y usos gastronómicos*, Lima: Ministerio de Agricultura (MINAG).

Ministry of Finance (2012) *Economic Survey 2011–12*, Government of India, New Delhi [online]. Available from www.indiabudget.nic.in/es2011-12/echap-08.pdf [accessed 19 November 2012].

Monteros, C., Yumisaca, F., Andrade-Piedra, J. and Reinoso, I. (eds) (2011) *Papas Nativas de la Sierra Centro y Norte de Ecuador: catalogo etnobotánica, morfológico, agronómico y calidad*, Quito: Instituto Nacional Autónomo de Investigaciones Agropecuarios (INIAP), Centro Internacional de la Papa (CIP) [online]. Available from www.scribd.com/fullscreen/77440828?access_key=key-1vzly24bijg5pb3k0anl [accessed 3 December 2012].

Mooney, P.R. (1983) 'The Law of the Seed: Another Development and Plant Genetic Resources', *Development Dialogue*, 1–2, Dag Hammarskjöld Foundation.

Moreno, J.D., del Socorro Cerón, M. and Valbuena, R.I. (2009) *Papas Nativas Colombianas: catalogo de 60 variedades*, Bogota: Corporación Colombiana de Investigación Agropecuaria (CORPOICA).

Mustafa, Y., Grando, S. and Ceccarelli, S. (2006) *Assessing the Benefits and Costs of Participatory and Conventional Barley Breeding Programs in Syria*, Aleppo, Syria: ICARDA.

Nishikawa, Y. (2001) 'New relations between gene bank and farmers in practical utilization of land-races: a case of Hiroshima Agricultural Gene Bank', *Journal of Agricultural Development Studies*, 12(1), pp. 76–83 (in Japanese, with English abstract).

Nishikawa, Y. (2011) 'Promotion of plant genetic resource utilization through Hiroshima Agricultural Gene Bank and farmers', *Reports on Production, Trade and Utilization of Seeds and Seedlings for Organic Agriculture*, 3, Japan Organic Agriculture Association (in Japanese).

Oscarsson, M., Anderson, R., Salomonsson, A.C. and Aman, P. (1996) 'Chemical composition of barley samples focusing on dietary fibre components', *Journal of Cereal Science*, 24(2), pp. 161–170.

Pelegrina, W. and Borja, P.P. (eds) (2002) *Frequently Asked Questions about Community Registry*, Quezon City: SEARICE.

Pistorius, R., Lim, E.S., Ghijsen, H. and Visser, B. (2009) *Results of an Online conference on 'Options for Farmers' Rights'*, IT/GB-3/09/Inf. 6 Add.2, Rome: FAO.

Protection of Plant Varieties and Farmers' Rights Authority (2011): Annual Report 2010–2011, New Delhi: PPVFR Authority.

Ramanna, A. (2006) *Farmers' Rights in India. A Case Study*, FNI Report 6/2006, Lysaker, Norway: Fridtjof Nansen Institute.

Ramanna, A. and Smale, M. (2004): 'Rights and Access to Plant Genetic Resources Under India's New Law', *Development Policy Review*, 22(4), pp. 423–42.

Rana, R.B., Garforth, C.J., Stahpit, B.R. and Jarvis, D.I. (2011) 'Farmers' rice seed selection and supply system in Nepal: understanding a critical process for conserving crop diversity', *International Journal of AgriScience*, 1(5), pp. 258–74.

Rhoades, R. and Booth, R. (1982) 'Farmer-back-to-farmer: a model for generating acceptable agricultural technology', *Agricultural Administration*, 11(2), pp. 127–37.

Robinson, R.A. (1996) *Return to Resistance. Breeding Crops to Reduce Pesticide Dependence*, Ottawa: International Development Research Centre.

Ruiz, M. and Vernooy, R. (eds) (2012) *The Custodians of Biodiversity: Sharing Access to and benefits of Genetic Resources*, Abingdon: Earthscan.

Santili, J. (2012) *Agrobiodiversity and the Law. Regulating Genetic Resources, Food Security and Cultural Diversity*, Abingdon: Earthscan.

Scurrah, M., Andersen, R. and Winge, T. (2008) *Farmers' Rights in Peru: Farmers' Perspectives*, FNI Report 16/2008, Lysaker, Norway: Fridtjof Nansen Institute.

Seshia, S. (2001) *Plant Variety Protection and Farmers' Rights in India: Law-making and the Cultivation of Varietal Control*, MPhil Dissertation, University of Sussex, UK: Institute of Development Studies.

Stapleton, P. (2006) 'Tying the genome up in knots', *Geneflow*, Rome: Bioversity International, p. 51.

Suneson, C.A. (1956) 'An evolutionary plant breeding method', *Agronomy Journal*, 48, pp. 188–91.

Terrazas, F. and Cadima, X. (2008) *Catálogo Etnobotánico de Papas Nativas: tradición y cultura de los ayllus del norte Potosí y Oruro*, Cochabamba: Fundación PROINPA.

Tripp, R., Louwaars, N., Van der Burg, W.J., Virk, D.S and Witcombe, J.R. (1997) 'Alternatives for seed regulatory reform: an analysis of variety testing, variety regulation and seed quality control', *Agricultural Research & Extension Network AgREN Papers*, 69, January 1997.

Ugarte, M.L. and Iriarte, V. (2000) *Papas Bolivianas: catalogo de cien variedades nativas*, Cochabamba: Fundación PROINPA.

Ugas, R. (ed.) (2008) *Pampacorral: catalogo de sus papas nativas*, Lima: Universidad Nacional Agraria la Molina (UNALM).

United Nations (2009) *The right to food. Seed policies and the right to food: enhancing agrobiodiversity and encouraging innovation*, Report of the Special Rapporteur on the Right to Food, A/64/170, New York: United Nations.

Urton, G. (2003) *Signs of the Inka Khipu: Binary Coding in Knotted-string Records*, Austin, Texas: University of Texas Press.

Walker, T.S. (2006): *Participatory varietal selection, participatory plant breeding, and varietal change*, Background paper for the World Development Report 2008, Washington D.C: The World Bank.

Winge, T. (2012) *A Guide to EU Legislation on the Marketing of Seed and Plant Propagating Material in the Context of Agricultural Biodiversity*, FNI Report 11/2012. Lysaker, Norway: Fridtjof Nansen Institute.

Wilson, E.O. (1992) *The Diversity of Life*, Cambridge, MA: Belknap Press of Harvard University Press.

Wolfe, M.S. (2000) 'Crop strength through diversity', *Nature*, 406, pp. 681–82.

Index

For Product Safety Concerns and Information please contact our EU
representative GPSR@taylorandfrancis.com Taylor & Francis Verlag GmbH,
Kaufingerstraße 24, 80331 München, Germany

Printed and bound by CPI Group (UK) Ltd, Croydon, CR0 4YY
08/12/2025
02014057-0004